An Introduction to Hard Ferrites
From Fundamentals to Practical Applications

Edited by

Gagan Kumar Bhargava[1], Pankaj Sharma[2], Sumit Bhardwaj[1], Indu Sharma[3]

[1] Department of Physics, Chandigarh University, Gharuan, Mohali, Punjab, India

[2] Applied Science Department, National Institute of Technical Teachers Training & Research, Chandigarh, India

[3] Department of Physics, Career Point University, Himachal Pradesh, India

Published by **Materials Research Forum LLC**
Millersville, PA 17551, USA

Published as part of the book series
Materials Research Foundations
Volume 142 (2023)
ISSN 2471-8890 (Print)
ISSN 2471-8904 (Online)

Print ISBN 978-1-64490-230-1
eBook ISBN 978-1-64490-231-8

This book contains information obtained from authentic and highly regarded sources. Reasonable efforts have been made to publish reliable data and information, but the author and publisher cannot assume responsibility for the validity of all materials or the consequences of their use. The authors and publishers have attempted to trace the copyright holders of all material reproduced in this publication and apologize to copyright holders if permission to publish in this form has not been obtained. If any copyright material has not been acknowledged please write and let us know so we may rectify this in any future reprints.

Distributed worldwide by

Materials Research Forum LLC
105 Springdale Lane
Millersville, PA 17551
USA
https://www.mrforum.com

Manufactured in the United States of America
10 9 8 7 6 5 4 3 2 1

Table of Contents

Preface

Hard ferrites have long been recognized as essential ingredients in the training of students while considering scientific and engineering applications owing to their excellent magnetic characteristics, which make them useful in various research applications such as permanent magnets, storage devices, and recording media. The reader must gain knowledge of the fundamental concepts of hard ferrites and the capacity to apply these principles in this field.

Books on hard ferrites generally cover very specialized parts and lack the basics and advanced processing of hard ferrites simultaneously. Keeping this in view, an attempt has been made to put things in an order that should be useful for students from postgraduate to research level and some extent, for established scientists.

Here, in this book, "An Introduction to Hard Ferrites: From Fundamentals to Practical Applications" an emphasis is laid on the fundamentals of hard ferrites. This has been followed by advanced synthesis and processing techniques for designing and developing various types of hard ferrites. The characterization part of hard ferrites is an integral part and is described wherever necessary.

The book is divided into seven chapters. The first chapter gives an overview of hard ferrites, their types, and their structures. Chapter two includes the advances in the processing of hard ferrites. Two chapters (3rd and 4th) are specifically designed for the case studies of hard ferrites to understand the concept of ferrites. The case studies include the electric and magnetic properties of $BaFe_{12}O_{19}$ and $SrFe_{12}O_{19}$ hard ferrites. Three chapters (5th to 7th) are there to address the need for specific applications of hard ferrites corresponding to permanent magnets, antenna and radar applications, and memory devices.

We hope the reader likes reading the book; we would greatly appreciate suggestions for the future volume.

Chapter 1

An Overview of Hard Ferrites: Types and Structures

Rohit Jasrotia[1,2*], Suman[3], Ankit Verma[4*], Rahul Kalia[1], Himanshi[1], Rajat Kaushal[1], Jyoti Prakash[1], Sachin Kumar Godara[5], and Pooja Puri[6]

[1] School of Physics and Materials Science, Shoolini University, Bajhol, Solan, H.P., India

[2] Himalayan Centre of Excellence in Nanotechnology, Shoolini University, Bajhol, Solan, H.P., India

[3] Department of Mathematics, School of Basic and Applied Sciences, Maharaja Agrasen University, Baddi, Himachal Pradesh, India

[4] Faculty of Science and Technology, ICFAI University, Himachal Pradesh, India

[5] Department of Apparel and Textile Technology, Guru Nanak Dev University, Amritsar

[6] Department of Chemistry, Bahra University, Waknaghat, Solan, H.P., India

rohitsinghjasrotia4444@gmail.com, ankitvermaarki26@gmail.com

Abstract

In 1950s, after the discovery of hexagonal ferrites called as hexaferrites, there has been an increasing level of curiosity in hexaferrites and still, it is increasing exponentially. Commercially and technologically, hexaferrites belong to a category of important magnetic materials due to their utilization in various applications such as permanent magnets, recording media, storage devices, rod antennas, high frequency devices, microwave devices etc. There are six types of hexagonal ferrites as M-type, Z-type, Y-type, X-type, U-type and W-type respectively, out of which, the most worldwide manufactured hexagonal ferrite is Barium M-type hexagonal ferrite named as BaM [1]. All categories of hexaferrites shows ferrimagnetic behaviour as their magnetic nature is inherently related to their crystal structure. The current chapter focuses on the classification and structure of hexagonal ferrites. In this book chapter, firstly, we report the six categories of hexaferrites as M, Y, W, X, Z, W and then, we provide a detailed comprehensive review of two approaches called as spinel and S/R/T based model for reporting the structure of hexagonal ferrites respectively. After that, we reported a comprehensive structure of each hexaferrite. Lastly, the concluding remarks have been presented in the current chapter.

An Introduction to Hard Ferrites: From Fundamentals to Practical Applications   Materials Research Forum LLC
Materials Research Foundations **142** (2023) 1-34   https://doi.org/10.21741/9781644902318-1

Keywords

Hexaferrites, Classification, Structure, Spinel Model, S/R/T Model

List of Abbreviations

PM	Permanent magnets
RM	Recording media
SD	Storage devices
MD	Microwave devices
HFD	High frequency devices
DC	Dielectric constant
FMR	Ferromagnetic resonance
GHz	Giga Hertz
SC	Satellites communications
LAN	Local area network
RST	Remote sensing
MA	Microwave absorbers
RAM	Radar absorbing materials
EMI	Electromagnetic interference
SD	Storage devices

Contents

1.   Introduction

In the last few decades, ferrites become important magnetic materials due to their superfluous magnetic characteristics which makes them useful for a number of research applications such as permanent magnets, storage devices, recording media etc. [1–10,10]. In 1950, according to the statement of Richard Feynman, ferrites are very difficult to investigate as a theoretical study but, as a part of technology and utilization, it is considered as one of the superior issues to analyse [11]. The first permanent magnet was found in 1951 and moreover, it was synthesized on the phenomenon of barium M-type hexaferrite usually named as BaM hexaferrite [12,13]. The researchers as well as scientists have paid great attention towards the fabrication of ferrites and hexaferrites for various applications in a well-defined way after their discovery in the 1950s [14]. Therefore, in 1955, the hexaferrites were synthesized for gyromagnetic applications in a defined manner [15,16]. The hexagonal ferrites are now the subject of a great deal of experimental investigation in the laboratory. [17]. Hexaferrites are fabricated for commercial use as well as technology use in various research areas. The hexagonal ferrites are classified into six different categories as M-type, Z-type, Y-type, W-type, X-type and U-type [18,19]. The magnetic characteristics of hexaferrites are inherently related to their crystal structure having crystalline behaviour, therefore, they show ferrimagnetic behaviour respectively and also, it indicates induced magnetization with a preferable orientation within the hexagonal based crystal structure [14,20]. As all the hexaferrites show magneto crystalline anisotropy, thus, on this phenomenon, they have been categorized further into two groups as uniaxial and ferroplana/hexaplana hexaferrites. The uniaxial hexaferrites have an easy axis of magnetisation whereas, the ferroplana/hexaplana are those hexaferrites with an easy plane of magnetisation [21,22]. The number of publications under the category of hexagonal ferrites from 1991-2020 is presented in figure 1. The current book chapter concentrate on the brief description of hexagonal ferrites especially, containing divalent metal cations as barium and strontium along their classification and structure of each of hexaferrites respectively.

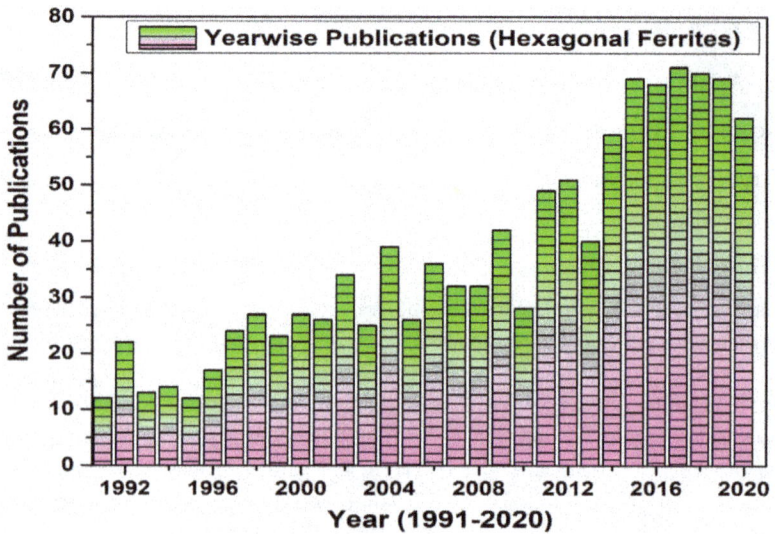

Figure 1: Year wise publications under the category of hexagonal ferrites from 1991 to 2020 (Using Scopus Database)

2.   Classification and characteristics of hexaferrites

In 1925, the first magnetic mineral called magnetoplumbite was explained and therefore, the hexagonal crystal structure was derived with the chemical formula, $PbFe_{7.5}Mn_{3.5}Al_{0.5}Ti_{0.5}O_{19}$ in the year 1938 respectively. The pure form of magnetoplumbite mineral was found to be PbM or $PbFe_{12}O_{19}$ [23]. At the time of the Second World War, various isomorphous compounds-based materials along with an inclusion of BaM were proposed but they were not investigated structurally till that time. BaM or $BaFe_{12}O_{19}$ called as barium hexaferrite having hexagonal like crystal structure was reported by Went et al.

[13,24]. In 1950s, at the Philips laboratories, a comprehensive investigation of all the major phases of hexagonal ferrites were published, concluding in a book named as "Ferrites" by Smit and Wijn's in 1959 [25]. All these compounds comprise of a hexagonal like structure with two crystal lattice parameters called "a" and "c" representing thickness of hexagonal plane and altitude of a crystal. The hexagonal ferrites have been categorized as uniaxial and ferroplana/hexaplana as provided in table 1. Moreover, table 1 provides a comprehensive study of hexagonal ferrites about their types, chemical composition and characteristics. The hexagonal ferrites are categorized into six types as listed below [26].

- o M-type hexagonal ferrite (BaM)
- o Z-type hexagonal ferrite (Co_2Z)
- o Y-type hexagonal ferrite (Co_2Y)
- o W-type hexagonal ferrite (Co_2W)
- o X-type hexagonal ferrite (Co_2X)
- o U-type hexagonal ferrite (Co_2U)

All these types of hexagonal ferrites can be fabricated for commercial or technological use with the utilization of various synthesis methods such as sol-gel, auto-combustion, co-precitation, hydrothermal, citrate precursor, solid-state or conventional ceramic, microemulsion, green synthesis [7] and many more [8,10,27]. The different categories of hexagonal ferrites have been explained as given below.

Table 1: Types, chemical composition and characteristics of hexagonal ferrites [2,26]

Hexaferrite	Chemical Composition	Room Temperature Magnetisation
M-type	$BaM/BaFe_{12}O_{19}/SrFe_{12}O_{19}$	Uniaxial
Z-type	$Co_2Z/Ba_3Co_2Fe_{24}O_{41}/Ba_3Me_2Fe_{24}O_{41}$	In plane
Y-type	$Co_2Y/Ba_2Co_2Fe_{12}O_{22}/ Ba_2Me_2Fe_{12}O_{22}$	In plane
W-type	$Co_2 Ba_2Co_2Fe_{12}O_{22}/Ba_2Me_2Fe_{12}O_{22}$	In cone
X-type	$Co_2X/Ba_2Co_2Fe_{28}O_{46}/Ba_2Me_2Fe_{28}O_{46}$	In cone
U-type	$Co_2U/Ba_4Co_2Fe_{36}O_{60}/Ba_4Me_2Fe_{36}O_{60}$	In plane

2.1   M-type hexaferrite

The barium M-type hexagonal ferrite called as BaM was firstly confirmed in the year 1936 and moreover, at that time, their crystal structure was unconfirmed. Initially, the barium hexaferrite was called as ferroxdure in order to differentiate it from the spinel-based ferrite. The M-type hexagonal ferrite have the chemical composition $MeFe_{12}O_{19}$, where Me is any divalent metal cation as Ba^{2+}, Sr^{2+} or Pb^{2+}, etc. with a high uniaxial room temperature magnetisation along the c-axis [28]. The molecular weight and hardness of barium M-type hexaferrite is found to be 1112 g and 5.9 GPa respectively [29,30]. Also, the strontium M-type hexaferrite called as SrM has a molecular weight of 1062 g in which the Sr^{2+} ions replaced the Ba^{2+} ions in the crystal structure of $BaFe_{12}O_{19}$ hexaferrite [18]. The physical characteristics of BaM shows similar behaviour to that of the characteristics of SrM hexaferrite respectively [21]. The physical characteristics of BaM and SrM Hexagonal ferrite provided in table 2.

Table 2: Physical parameters of BaM and SrM hexaferrite [2]

Hexaferrite	Chemical Composition	Molecular Weight (g)	Density (g/cm^3)	Melting Point (Kelvin)	Lattice Parameters (a &c)
BaM	$BaFe_{12}O_{19}$	1112	5.295	1611	a = 5.888, c = 23.189
SrM	$SrFe_{12}O_{19}$	1062	5.101	1692	a = 5.884, c = 23.063

2.2   Z-type hexaferrite

When the Y-type ferroxolana hexaferrites were discovered, the Z-type hexaferrites with the chemical composition $Ba_3Me_2Fe_{24}O_{41}$ were discovered at the same time. Where 'Me' represents the divalent metal cations, and moreover, it has parallel uniaxial anisotropy corresponding to the c-axis and a-axis, respectively [31]. At low temperatures, Z-type hexagonal ferrite exhibits a simple cone magnetization, with a 65° magnetization angle matching to the c-axis. The preferred magnetisation remains in the basal plane until between 207 and 242 degrees Celsius, when it switches to the c-axis. [25, 32]. Z-type hexagonal ferrite has the following physical characteristics, which are listed in Table 3.

Table 3: Physical parameters of Z-type hexaferrite [31]

Hexaferrite	Chemical Composition	Molecular Weight (g)	Density (g/cm^3)	Melting Point (Kelvin)	Lattice Parameters (a & c)
Z-type	$Co_2Z/Ba_3Me_2Fe_{24}O_{41}$	2522	5.35	1611	a = 5.91, c = 52.30

2.3   Y-type hexaferrite

The Y-type hexagonal ferrite called as Co_2Y having the chemical formula $Ba_2Me_2Fe_{12}O_{22}$ was the first ferroxplana hexaferrite ever discovered, where Me indicates the small divalent metal cation [33]. At room temperature, all the Co_2Y hexaferrites have a preferable plane of magnetisation at 90° angle corresponding to the hexagonal c-axis respectively and also, out of all these Co_2Y hexaferrites, Cu_2Y is the only hexaferrite with a preferable uniaxial direction of magnetisation [34]. In addition, below -58 C°, the room temperature planar magnetic anisotropy of Y-type hexaferrites has been converted to a cone of magnetisation. Table 4 indicates the physical parameters of Y-type hexagonal ferrite.

Table 4: Physical parameters of Y-type hexaferrite [19,33]

Hexaferrite	Chemical Composition	Molecular Weight (g)	Density (g/cm³)	Lattice Parameters (a & c)
Y-type	Co_2Y/ $Ba_2Me_2Fe_{12}O_{22}$	1410	5.40	a = 5.88, c = 43.50

2.4   W-type hexaferrite

The W-type hexagonal ferrites with the chemical composition $BaMe_2Fe_{16}O_{27}$, were found as an assorted phase of both M-type and X-type hexaferrites, where 'Me' indicates transition divalent metal cations. FeW called as $BaFe_2Fe_{16}O_{27}$ was the first reported W-type hexaferrite having an easy axis of magnetisation along the c-axis of hexagonally based lattice structure and moreover, all W-type hexaferrites except Co_2W have a uniaxial anisotropy respectively. Table 5 represents the physical parameters of W-type hexaferrite [21,35].

Table 5: Physical parameters of W-type hexaferrite [21,35]

Hexaferrite	Chemical Composition	Molecular Weight (g)	Density (g/cm³)	Lattice Parameters (a & c)
W-type	Co_2W/ $BaMe_2Fe_{16}O_{27}$	1581	5.31	a = 5.83, c = 32.84

2.5   X-type hexaferrite

A uniaxial anisotropy along the c-axis of the hexagonally based crystal structure was seen in 1952 in the first known X-type hexagonal ferrite, Fe2X, which was discovered in 1952. For example, the chemical composition $Ba_2Me_2Fe_{28}O_{46}$ denotes the first-row transition divalent metal cations, whereas 'Me' denotes the second-row transition divalent metal cations. It has been observed as a mixed phase of M-type/W-type hexaferrite with room temperature uniaxial magnetic anisotropy, with the exception of Co_2X. A cone of magnetisation is seen in both directions along the c-axis of Co_2X hexaferrite when viewed at an angle of 74°. As shown in Table 6, the physical characteristics of X-type hexaferrite are given.

Table 6: Physical parameters of X-type hexaferrite [21]

Hexaferrite	Chemical Composition	Molecular Weight (g)	Density (g/cm³)	Lattice Parameters (a & c)
U-type	$Co_2U/$ $Ba_2Me_2Fe_{28}O_{46}$	2688	5.29	a = 5.889, c = 86.303

2.6   U-type hexaferrite

The U-type hexagonal ferrite with the chemical composition, $Ba_4Me_2Fe_{36}O_{60}$, where 'Me' represents the first-row transition metal cations have uniaxial anisotropy. Except all U-type hexaferrites, Co_2U have room temperature planar anisotropy respectively. The table 7 indicates the physical parameters of U-type hexagonal ferrites.

Table 7: Physical parameters of U-type hexaferrite [21]

Hexaferrite	Chemical Composition	Molecular Weight (g)	Density (g/cm³)	Lattice Parameters (a & c)
U-type	$Co_2U/ Ba_4Me_2Fe_{36}O_{60}$	3622	5.01	a =5.89, c = 113.22

3.   A brief description of the solid-state chemistry of hexaferrites

The solid-state reactions of the $BaO.Fe3O3.MeO$ system have been extensively investigated utilising a conventional ceramic synthesis method based on barium carbonate and barium oxides as starting materials. Using the Vinnik-created phase diagrams (see Figure 2), the x-ray spacings for specimens of barium hexaferrites containing more than 50% of $Fe2O3$ were calculated for the M-type, Y-type, Z-type, and W-type varieties of the material. Neither $Co2X$ nor $Co2U$ hexaferrites could be identified in polycrystalline specimens after two hours of calcination at 1200 °C followed by four hours of annealing at 1250 °C [36]. Additionally, the existence of a little amount of variation in the values of the lattice parameters (a & c) for specimens of phase diagrams of both W- and Z-type hexagonal hexaferrites shows that the M, Y, and W-types may absorb in Z-type hexaferrite. Additionally, it was revealed that M-type hexaferrite absorbs more effectively in W-type hexaferrite than in Y- or Z-typehexaferrites.

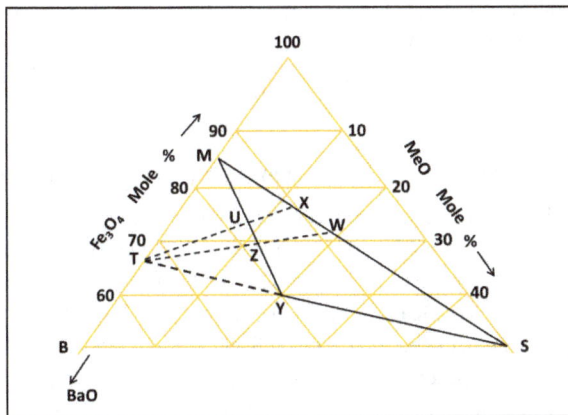

Figure 2: Schematic scheme of BaO–Fe₂O₃–MeO system [21]

Because the other phases dissolve considerably more slowly, dissolving the $BaFe2O4$ in HCl may remove the non-magnetic hexagonal ferrite from the mixture [37]. It was observed that producing a single-phase specimen of polycrystalline Y-type, Z-type, or W-type hexaferrite is very difficult, and the original specimen required to be reduced to a single phase by grain grinding. A flux approach was used to synthesise a high melting point hexaferrite hexamer consisting of Co2W/Co2Z/Co2X/Co2U hexaferrites, however single crystal growth of Co2Y and Fe2W proved challenging. All hexaferrites, according to Neckenburger and Severin's investigation, overlap at the same temperature [24, 38]. Hexaferrite phases of the U- and X-type are often unclear against this backdrop of mixed phases. Due to the comparable x-ray diffraction patterns and magnetic characteristics of hexagonal ferrites, hexaferrites contain similar basic building blocks inside their hexagonal structures. The Y-type, Z-type, and W-type phases have ceased to exist at temperatures greater than 1300 °C. This is due to the transformation of ferric ions into ferrous ions. Lotgering combined oriented grains of barium M-type hexaferrite with non-oriented grains of other non-magnetic materials to create oriented polycrystalline crystals with hexagonal shapes. This reaction produced orientated W, Y, and Z hexaferrites, as well as substituted M-type hexaferrites [39]. The co-precipitated precursors have been widely investigated for their use in the synthesis of M, Y, Z, and W hexaferrites. The nitrate salts were agitated until a concentration was obtained in a solution of ammonium oxalate. At this time, precipitation occurred, resulting in a mixture of partially degraded oxalates. They were then completely decomposed at 500°C, calcined at 800°C, and then burnt at 1100-1400°C. Around 650°C, the spinel phase starts to form. The BaM phase is formed around 800°C. Sudakar et al. investigated the BaO–NiO–Fe2O3 system between 80 and 90°C. At 750°C, the nanopowder precursors all contained the required hexaferrite phase in the single-phase hexagonal hexaferrite crystal structure [40].

4.   Approaches for better understanding of crystal structure of hexaferrites

The structure of six types of hexaferrites can be explained on the phenomenon of two types of approaches called as spinel-based model and S/R/T blocks-based model [2]. The magnetic mineral magnetoplumbite was first described in the 1920s. It was described as a hexagonal crystal of the empirical formula, $PbFe_{7.5}Mn_{3.5}Al_{0.5}O_{19}$ [21]. Philips Laboratories conducted many research in the 1950s to replicate a synthetic version of this mineral ($PbFe_{12}O_{19}$) and its analogs, for starting the development of the young field of hexaferrite. In these synthetic hexaferrites, lead was substituted for large divalent cations, namely strontium ($SrFe_{12}O_{19}$) and barium ($BaFe_{12}O_{19}$). At a chemical level, hexaferrite can be understood as the combination of spinel ferrites, Y-type/M-type hexaferrites. For instance, the Z-type hexaferrites may be represented as the sum of the Y-type and M-type hexaferrites. Crystallographically, all hexaferrites can be understood as a composition of spinel and hexagonal layers. The unit cell of each subgroup of hexaferrite would be created by different stacking arrangements of these basic components. Therefore, the family of hexaferrites do not differ in their crystallographic framework. This wide range of stacking possibilities has a massive effect on their magnetic properties.

4.1   Spinel based model

The hexaferrites have a crystal structure that is similar to that of spinel ferrites, consisting of plates of cubic densely packed oxygen anions interspersed with lesser sized metal cations. They are found in the interstices of the metal cation complex (tetrahedral site-A) and (octahedral site-B). Additionally, the spinel crystal structure is determined by the four oxygen atom-based layers that repeat the three vertical levels, forming an ABCABC crystal lattice between each layer. The interstices between the layers are filled by one B-site/three B-sites and two A-sites. Hexaferrites have a crystal structure composed of two plates, designated S_4 and S_6, which are joined together by layers (B_1 and B_2) containing two barium atoms. As a result, the M-type, Y-type, and Z-type hexaferrites are composed of a single plate designated as S_4, the X-type is composed of two plates designated as S_4 and

S_6, and the W-type hexaferrite is composed of a single plate designated as S_6. In contrast, the B_1 layer acts as a single hexagonal layer, whereas the B_2 layer behaves as two hexagonal layers. Additionally, the B_1 layer is found in M, W, and Z-type hexaferrites, which contain one barium atom in the location of 4 oxygen anions. On the other hand, the B_2 layer is found in Z-type hexaferrites, which contain two Ba atoms in the location of 8 oxygen anions, and the B_3 layer is found in M-type hexaferrites, which contain 2 Ba atoms in By combining the S and B layers, which are referred to as S_2, M_5 and Y_6 units, the crystal structure of hexaferrites may be investigated in great detail. The S_2 layer represents the spinel layers, the M_5 layer represents B_1 inserted between the four spinel layers, and the Y_6 layer represents two B2 layers placed between the four spinel layers, as shown in Figure 3 [2].

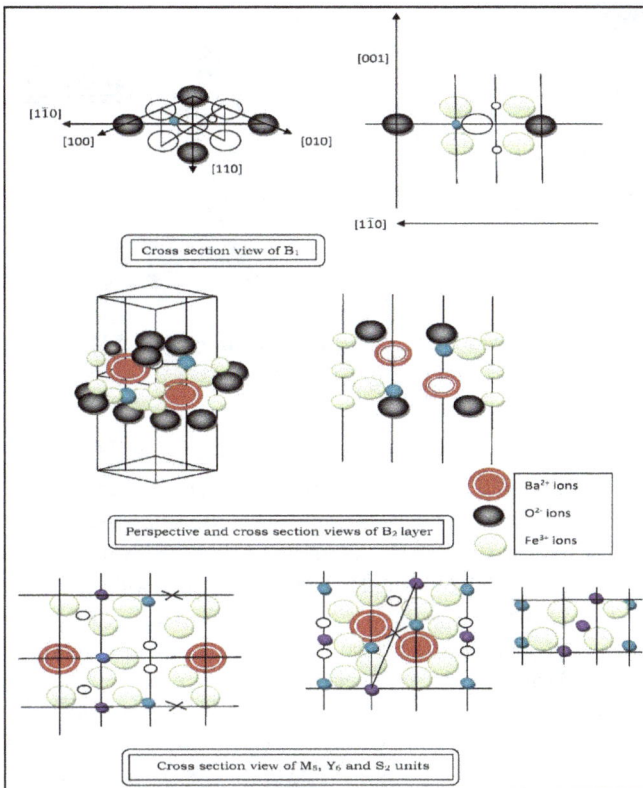

Figure 3: Cross-sectional interpretation of B_1 layer, B_2 layer and M_5, Y_6 and S_2 units [2]

4.2   S/R/T blocks based model

All known hexagonal ferrites are composed of structural blocks S, R, and T, and their crystalline structure is a superposition of these three blocks. When the letters S, R, and T are rotated 180 degrees around the hexagonal c-axis, they become the letters S, R, and T. The viewpoint, space-filling, ball, and sticky views of S/R/T blocks in hexaferrite crystal structure are shown in Figures 4a and 4b (b).

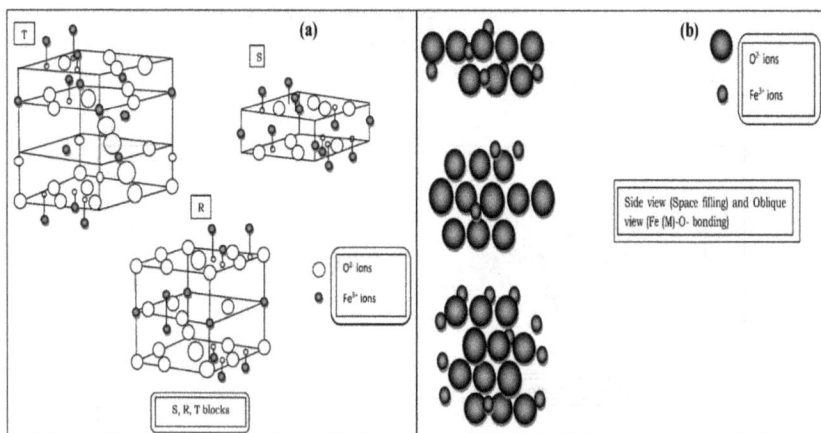

Figure 4:(a) Perspective(b) Space-filling,ball and sticky view of S/R/T blocks in the crystal structure of hexaferrites [21]

The repeating unit 'S' is composed of $[Me_2^{2+}Fe_4^{3+}O_8]^0$ (S^0) or $[Fe_6^{3+}O_8]^{2+}$ (S^{2+}), each of which has an uncompensated or neutral charge of +2. The 'R' subunit is composed of $[Me^{2+}Fe_6^{3+}O_{11}]^{2-}$, while the 'T' unit is composed of $[Ba_2^{2+}Fe_8^{3+}O_{14}]^0$. The subunit 'R' interacts with the subunit 'S^{2+}' to form the neutral block (RS), which has the chemical formula $MeFe_{12}O_{19}$ (M-phase). Similar, the 'T' subunit interacts with the S0 to generate the neutral block (TS), which has the chemical formula $Ba_2Me_2Fe_{12}O_{22}$ (Y-phase). Other stacking sequences of cubic and hexagonal basic units are also known [41–43], resulting in a variety of hexaferrite compositions including W, X, Y, Z, and U-type hexaferrite.

The repeating unit 'S' is made up of either $[Me_2^{2+}Fe_4^{3+}O_8]^0$ (S^0) or $[Fe_6^{3+}O_8]^{2+}$ (S^{2+}), both of which have an uncompensated or neutral charge of +2. The 'R' subunit is made up of $[Me^{2+}Fe_6^{3+}O_{11}]^2$, while the 'T' subunit is made up of $[Ba_2^{2+}Fe_8^{3+}O_{14}]^0$. The subunits 'R' and 'S^{2+}' combine to create the neutral block (RS), which has the chemical formula $MeFe_{12}O_{19}$

(M-phase). Similar to the previous example, the 'T' subunit interacts with the S^0 to produce the neutral block (TS), which has the chemical formula $Ba_2Me_2Fe_{12}O_{22}$ (Y-phase). Additionally, other stacking sequences of cubic and hexagonal basic units are known [41–43], resulting in a range of hexaferrite compositions, including W, X, Y, Z, and U-type hexaferrite.

Table 8: Basic properties of six types of Hexaferrites where *represents 180° rotation of block around the c-axis [2,21]

S. No.	Type of Hexaferrite	Chemical Formula	Spinel plate and hexagonal layers	S/R/T blocks Crystal structure
1	M	$BaFe_{12}O_{19}$	$2M_5=B_1S_4B_1S_4$	SRS*R*
2	W	$BaMe_2Fe_{16}O_{27}$	$2W_7=B_1S_6B_1S_6$	SSRS*S*R*
3	X	$Ba_2Me_2Fe_{28}O_{46}$	$3X_{12}=B_1S_4B_1S_6B_1S_4B_1S_6B_1S_4B_1S_6$	SSRS*S*R*
4	Y	$Ba_2Me_2Fe_{12}O_{22}$	$2Y_6=B_2S_4B_2S_4B_2S_4$	3(ST)
5	Z	$Ba_3Me_2Fe_{24}O_{41}$	$2Z_{11}=B_2S_4B_1S_4B_2S_4B_1S_4$	STSRS*T*S*R*
6	U	$Ba_4Me_2Fe_{36}O_{60}$	$U_{16}=B_1S_4B_2S_4B_1S_4$	SRS*R*S*T*

5. Crystal structure of hexaferrites

The hexagonal ferrite belongs to a class of hexaferrites having a hexagonal crystal structure along with a distinct c-axis. The composition of hexaferrites is provided in figure 5.

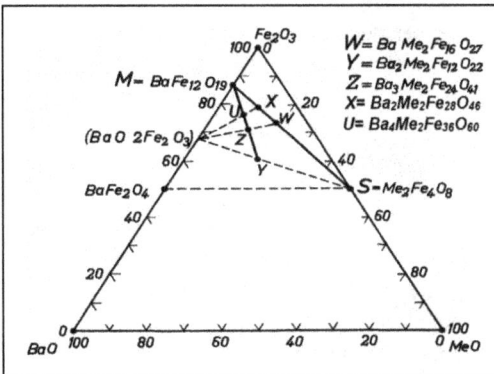

Figure 5: Schematic representation of composition of Hexaferrites(Copyright permission with the license number: 5146980447011) [19]

Hexaferrites' magnetic properties are related to their crystalline formations. Hexaferrites and hexaplana ferrites are two different forms of hexagonal ferrites having anisotropic magnetism. Hexaferrites have a typical composition of $MFe_{12}O_{19}$, where 'M' is a divalent metal cation such as Ba, Sr, or Pb [24, 44, 45]. While both hexaferrite and spinel ferrite have a similar crystal structure, with densely packed oxygen ions, the primary distinction is that certain layers in hexaferrite include metal ions with the same ionic radius as the oxygen ions. The metals occupy three distinct locations in the crystal structure (octahedral, trigonal bi-pyramid, and tetrahedral), which was further subdivided into five sublattice categories denoted as 12k, 2a, 2b, $4f_1$, and $4f_2$, respectively. Three sublattices, 12k, 2a, and 2b, have a high spin, while $4f_1$ and $4f_2$ have a low spin. There are six subtypes of hexagonal ferrites, denoted by the letters M, W, X, Y, Z, and U [1] [2] [5] [6], while (MO + MeO) and Fe_2O_3 are present in the proportions of 1:6, 3:8, 4:6, 4:14, 5:12, and 6:18 in the M, W, Y, X, Z, and U hexaferrites, respectively. 'Me' represent the transition cation (e.g., magnesium, zinc, manganese, cobalt, etc.) or a combination of cations comparable to spinel, where M represents the ions Ca, Sr, lead, barium, and lattice. Furthermore, other trivalent cations, such as Ga^{3+}, Al^{3+}, Sc^{3+}, and In^{3+}, may be employed to substitute Fe^{3+} ions in a variety of applications [46]. The dense packing of oxygen ion layers in hexagonal ferrite results in a crystalline structure that is hexagonal in shape. Substitutional entry of the heavy Ba or Sr ions into the oxygen layers is seen, while the divalent and trivalent metallic cations are detected in interstitial locations in the hexagonal crystal structure.

5.1 M-type hexaferrite

M-type hexaferrite was discovered initially and remains the most economically relevant hexagonal ferrite [47]. In the early 1950s, Phillips [21] was the first to explore and analyse it magnetically. M-type hexaferrite is a solid solution with the same structure as magnetoplumbite. Its chemical formula is MeO.6Fe2O3 or MeFe12O19, where Me may be divalent ions such as Ba2+, Sr2+, or Pb2+ [48,49]. It is composed of spinel blocks connected by a barium or strontium ion block R. Each spinel block has two oxygen layers, S and S*. On each side of the barium layer are two oxygen layers. The cross-sectional picture in Figure 6 shows M-type Hexaferrite. Between the layers containing the barium ion, four oxygen layers are packed in a cubical way. To avoid the barium ion's basal plane being identical to the R block, the blocks immediately before and following the R block must be rotated 180 degrees. A molecule is composed of five oxygen layers, whereas a unit cell is composed of two molecules. The crystal structure is denoted by RSR*S*, whereas the space group is denoted by P63/mmc [49]. The hexagonal structure of M-type hexaferrite crystallises with 64 ions per unit cell and 11 symmetry points. As a result, the unit cell consists of 38 oxygen ions, 24 ferric ions, and two metal cations (Me = Ba2+, Sr2+, Pb2+, and La3+). The 24 ferric ions are kept in six separate locations denoted by the designations 2a, 2b, 4f1, 4f2, and 12k. Sublattices 2a, 4f2, and 12k are octahedral, 4f1 is tetrahedral, and the last is tetrahedral, with the ferric ion surrounded by five oxygen atoms to produce the 2b site, a trigonal bipyramidal site. The cross-sectional picture in Figure 7 shows barium M-type hexaferrite. The arrows represent the direction in which Fe ions spin polarise. The unit cell contains 38 O2- ions, two Sr2+ ions, and 24 Fe3+ ions. The spins of Fe3+ ions at 12k, 2a, and 2b sites are up, while those in 4f1 and 4f2 sites are down, resulting in a net total of eight spins up and a total moment of (8*5) = 40B per unit cell containing two Sr2+ ions. It is composed of five distinct crystallographic orientations of strontium hexaferrite (SrFe12O19).

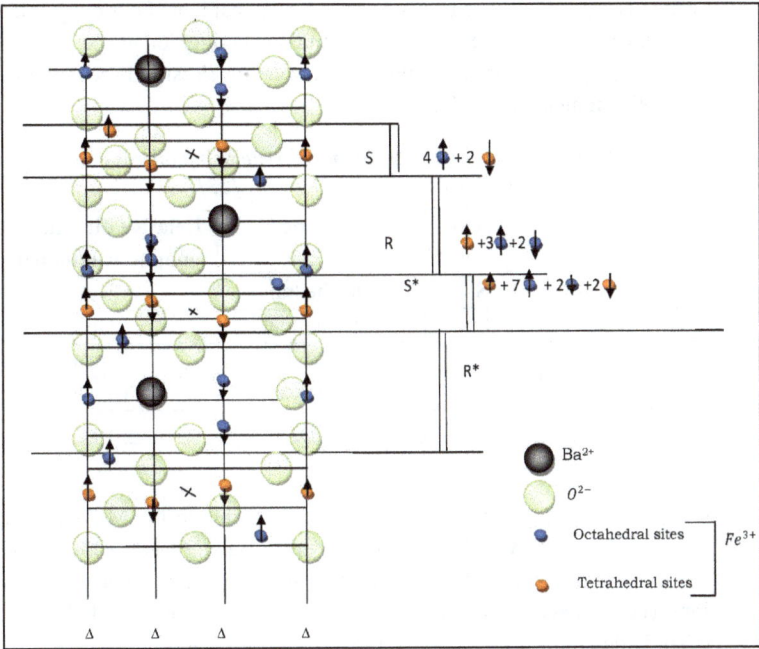

Figure 6: Cross-sectional view of structure of M-type hexaferrite [2]

This ferrimagnetic material's high saturation magnetism and coercivity are owed to its strong magneto-crystalline anisotropy field. Throughout the stacking of M-type hexagonal ferrite, layers of spinel (S = $Fe_6O_8^{2+}$) and hexagonal (R = $MFe_6O_{11}^{2-}$) ferrite alternated. The O^{2-} ions are layered densely, with the M^{2+} ions filling the hexagonal layer in lieu of an O^{2-}

ion. Three parallel (2a, 12k, and 2b) and two antiparallel (4f₁ and 4f₂) sublattices are connected through a super-exchange interaction with O^{2-} ions [51, 52]. In M-type hexaferrite, anisotropy appears in two forms: c-axis and c-plane. Simple magnetization along the c-axis and in the c-plane distinguishes each kind. Following the four oxygen ion levels, three oxygen ions and one Ba^{2+} ion are added, followed by the same sequence but with the Ba^{2+} ion from the preceding layer diametrically opposite the Ba^{2+} ion from the preceding layer; and finally, the same sequence but with the Sr^{2+} ion. In the interstices of these ten layers, Fe^{3+} ions are plentiful. On the other hand, the trigonal bi-pyramid site is unusual in the spinel structure due to the presence of five oxygen ions in the same layer as the Ba^{2+} ion. These are the triple symmetry axes, shown by spin orientation arrows. A primordial cell is composed of four layers of large hexagonal ions. Four oxygen ions are included in this layer, whereas three oxygen ions and one barium ion are contained in the fifth layer. The ionic sites in Barium M-type hexaferrite are classified according to their block and magnetic moment orientations in Table 9.

Table 9: Classification of ionic sites in Barium hexaferrite using blocks and the relative orientation of magnetic moments [53]

S. No.	Sites	Blocks	Site types	Relative direction of magnetic moments
1	Fe1	S	Octahedral	Up
2	Fe2	R	Bipyramidal	Up
3	Fe3	S*	Tetrahedral	Down
4	Fe4	R*	Octahedral	Down
5	Fe5	R/S	Octahedral	Up

The crystal structure of Ba and Sr hexaferrites is comparable to the M-type hexagonal ferrite. Only sintered magnets with higher coercivity than $BaFe_{12}O_{19}$ (SrM) have a similar crystalline structure and magnetic characteristics. A super-exchange interaction connects two iron ions (Fe^{3+}) magnetic moments aligned along the c-axis [53]. Iron-orange-ferromagnetism is determined by the angle produced between ferric cations and oxygen [25]. A single crystal's ferromagnetic resonance line width may range from 10 to 20 Oe. One use for M-type Strontium hexaferrite is magneto-optical. The DC electrical resistivity of Strontium hexaferrite is important for determining its applicability for microwave devices. Generally, high direct current resistivity materials are better for use in microwave devices. M-type hexaferrite is a commonly utilised material in permanent magnet manufacture, accounting for over 90% of yearly production due to its low cost and good magnetic properties. Aside from permanent magnets, permanent magnetic materials are used in a range of other items such as loudspeakers and moving coil instruments like galvanometers and ammeters. Magnetic motors are utilised in many applications, including automotive windshield wipers, heating fans, electric cutting blades, toothbrushes, and various home goods. Other uses for higher gigahertz microwaves include radar, stealth systems, satellite communication, precision navigation, and remote sensing. Microwave absorbers are in great demand for military applications. Putting a microwave absorbent coating on military planes and vehicles may help them evade detection. M-type hexaferrite is now commonly employed in the upper gigahertz range [54–56].

5.2   W-type hexaferrite

Philip's research laboratories discovered the W-type hexaferrite with the chemical formula $BaMe_2Fe_{16}O_{27}$ in 1980, where 'Me' represents transition divalent metal cations [36], which were later named after him. Figures 7(a) and 7(b) illustrate different orientations of W-type hexaferrites from a perspective and cross-sectional viewpoint, respectively (b). As is the case with M-type hexaferrite, it shows magnetic uniaxial anisotropy parallel to the axis of the hexagonal structure. Due to its moderate magnetic characteristics, chemical stability, cheap cost, strong coercivity, and high magnetic energy product, this hexaferrite has gained increasing attention in recent decades [57,58]. When compared to the M-type structure, the W-type structure is more similar to it since it contains one barium layer for every seventh oxygen layer inside the spinel structure. The saturation magnetization and Neel temperature of W-type hexaferrite may be tailored to the application by altering the divalent metal ions used.

This structure is a superposition of R and S blocks along the hexagonal c-axis (R = $BaFe_6O_{11}^{2-}$ hexagonal and S = $Fe_6O_8^{2+}$ cubic spinel). The O^{2-} ions are concentrated in thick layers, one Ba^{2+} ion is concentrated in the R block, and five Fe^{3+} ions are concentrated in the interstitial sublattices. The R block has one Ba^{2+} ion in place of an O^{2-} ion. Permanent magnets, bubble domain memories (including flash memory), high-frequency microwave devices (including satellites), and data processing systems all use W-type hexaferrite [59,60]. The method of manufacture, the sintering temperature, the kind and quantity of substitution all have an effect on the structural and magnetic characteristics of W-type hexaferrite. [61,62]. Spin reorientation transitions (SRTs) may occur in W-type ferrites $AMe_2Fe_{16}O_{27}$ (M = magnesium, manganese, Fe, Co, Ni, Cu, Zn) as a result of temperature or magnetic field changes [62–64]. Chemical reactions inside the mixture may cause the transition temperatures to fluctuate (substitution of divalent metal cation). Additionally, some SRT may be of the first order, meaning that W-type hexaferrites might be employed for refrigeration at ambient temperature [65]. There are three types of coordination: 12k (octahedral), 4e, 4fIV (tetrahedral), and 2d (bipyramidal) [66,67]. W-type hexaferrites bind cations through seven non-equivalent sublattices with varying amounts of R-blocks (including a Ba layer) and S-blocks [68] (containing a spinel layer). The coordination, spin orientation, block, and number of Fe ions at each W-type hexaferrite site are listed in Table 10.

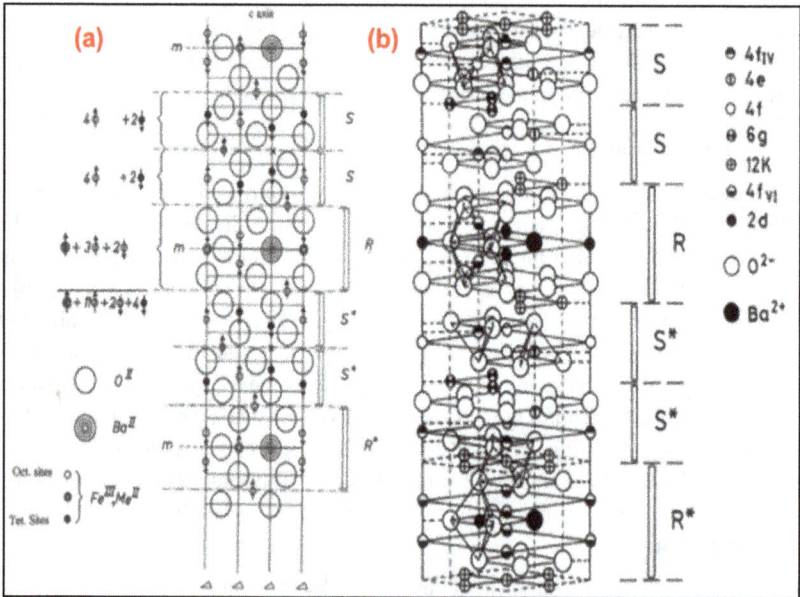

Figure 7: (a) Cross-sectional (b) Perspective view of the structure of W-type hexaferrite(Copyright permission with the license number: 5146980447011)[21]

The only five magnetically incompatible sub-lattices are as follows: The numbers 4e and IV create the fIV magnetic sub-lattice, while 6g and 4f form the 2b sub-lattice, also known as the b magnetic sublattice [69]. The cross-sectional and perspective architectures of W-type hexaferrite are shown in Figures 8(a) and 8(b), respectively (b). Numerous elements, including as manufacturing process, chemical composition, sintering temperature and duration, substitution type and amount, and others, all have an effect on the physical and magnetic properties of ferrites [70].

Table 10: Coordination, spin direction, block, and the number of Fe ions for each site of W-type hexaferrite [53]

S. No.	Magnetic site	Crystallographic site	Co-ordination	Spin	Block	Number
1	f_{VI}	$4f_{VI}$	Octahedral	Down	R	2
2	a	6g	Octahedral	Up	S-S	3
		4f		Up	S	2
3	f_{IV}	4e	Tetrahedral	Down	S	2
		$4f_{IV}$		Down	S	2
4	K	12k	Tetrahedral	Up	R-S	6
5	B	2d	Hexagonal	Up	R	1

5.3   X-type hexaferrite

Its chemical formula is $Ba_2Me_2Fe_{28}O_{46}$, where 'Me' stands for divalent metal ions. Hexaferrite is a hexavalent iron oxide crystal. In the first row of transition components, he found it 50 years ago. Real-space coordinates for oxygen and barium ions as well as transition metal ions were reported by Braun for the first time in $Ba_2Fe_{30}O_{46}$ [71]. Three-oxygen layer blocks R (BaFe6O11) and S (Fe6O8) are comprised of BaFe6O8 and BaFe6O11. There are a total of six R and S blocks stacked along the hexagonal c-axis in the model SRS*S*S*R. It's not uncommon for Type X phases (hexagonal ferrites) to be confused for M and W phases. By stacking R and S blocks along the hexagonal c-axis, the X-type hexaferrite is formed, as illustrated in table 11.

Table 11: Coordination, spin direction and block of X-type hexaferrite [53]

S.No.	Block	Coordination	Number per block	Expected spin direction
1	R	Octahedral	2	Down
2	R	Trigonal bipyramidal	1	Up
3	R-S	Octahedral	3	Up
4	S	Octahedral	1	Up
5	S	Tetrahedral	2	Down
6	S-S	Octahedral	3	Up

R and S blocks have three octahedral cation sites, as do S and S blocks. An S-block has two tetrahedral and one octahedral bipyramidal site, resulting in an elementary cell with three chemical formula units (Z = 3), with the following properties: 5.88, 84.11, and 5.30 g/cm^3.The unit cell in R$\overline{3}$m space symmetry [72,73] is made up of four layers of M- and W-type structures stacked hexagonally. Figure 8 shows natural X-type hexaferrites cross-sectioned.

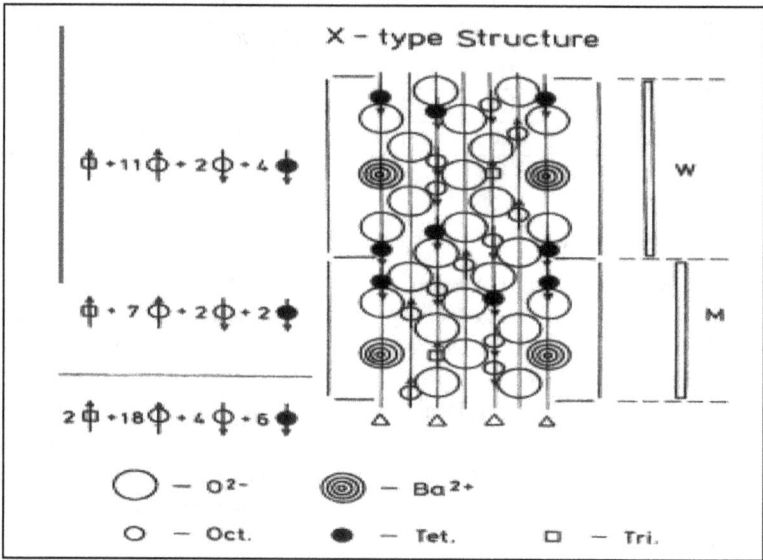

Figure 8: Cross-sectional view of structure of X-type Hexaferrite(Copyright permission with the license number: 5146980447011) [21]

5.4   Y-type hexaferrite

Hexaferrites are blocks with spinel architecture constructed of Ba-O alternately laminated. An ST stacking block model represents each spinel layer and Ba-O block three times in a hexagonal unit cell. Ba represents alkaline earth metals, whereas Me represents divalent ions like Mg or Zn. It's at the vertices of an octahedron or tetrahedron depending on the ion's structure. Antiferromagnetic exchange interactions occur between neighbouring Fe^{3+} ions, resulting in a complicated magnetic structure [53]. It is composed of two fundamental layers, the first of which contains just oxygen and the second of which contains an orderly replacement of barium for every fourth oxygen. Smaller cations (e.g., Me^{2+} or Fe^{3+}) present within the anion framework may be ignored in stacking relationships. The O and Ba-O layers interlock to create blocks. S and T are two-layer blocks. Next is the ST unit. It has 18 oxygen layers, 6 oxygen layer repetition distance, and 43.56 Å c-axis length.The hexagonal symmetry unit cell contains 18 oxygen layers and a six repetition distance. It has four big ions in each layer. There are four layers of four oxygen ions each, and then two layers of three oxygen ions and one barium ion. The unit cell is made up of three formula units in the sequence STSTST. On the six sublattices, the metallic cations are shown in Table 12.

Table 12: Number of ions per unit formula, coordination, and spin orientation for the various metallic sublattice of Y-type hexaferrite [53]

S. No.	Sublattice	Coordination	Block	Number of ions	Spin
1	$6c_{IV}$	Tetrahedral	S	2	Down
2	$3a_{VI}$	Octahedral	S	1	Up
3	$8h_{VI}$	Octahedral	S-T	6	Up
4	$6c_{VI}$	Octahedral	T	2	Down
5	$6c_{IV}$	Tetrahedral	T	2	Down
6	$3b_{VI}$	Octahedral	T	1	Up

The T block has three octahedral ions of sublattices 6cVI and 3bVI, with the centre 3bVI ion sharing two faces of coordination with each of the two 6cVI ions on each side. Due to the increased electrostatic repulsion between the cations in this arrangement, the structure's potential energy is greater; as a consequence, low-charge ions are more likely to choose such places. The multiferroic properties of Y-type hexagonal ferrite have aroused the interest of a significant number of researchers in recent years [74,75]. These hexaferrites are used in a wide range of applications, including electronic communication and microwave devices [76,77]. In electronic communication, low-permeability materials (0–10 GHz) are needed for ferrite device reduction, hence Y-type hexaferrite is being investigated as a suitable option [78–80]. The high permeability of Y-type hexaferrite is due to spin rotation and domain wall motions inside the material's crystal structure. Additionally, Y-type hexaferrite is a soft magnetic material that is often utilised in the VHF and UHF bands [53]. Figure 9(a) and 9(b) illustrate Y-type hexaferrites in cross-section and from a perspective, respectively.

5.5   Z-type hexaferrite

Z-type hexaferrite is one of the most advanced compounds in the hexaferrite family, boasting high permeability up to GHz, high resistivity, and great chemical stability, among other features. Z-hexaferrite is generated when M- (BaFe12O19) and Y- (Ba2Me2Fe12O22) type hexaferrites are combined. The STSR block-based model [81] depicts the crystal structure as being formed of 33 layers stacked along the hexagonal c-axis and classified into S, R, and T blocks, which correspond to the STSR block-based model's S, R, and T blocks. The Z-type hexaferrite is shown in Figures 10(a) and (b) in cross-sectional and viewpoint perspective views, respectively.

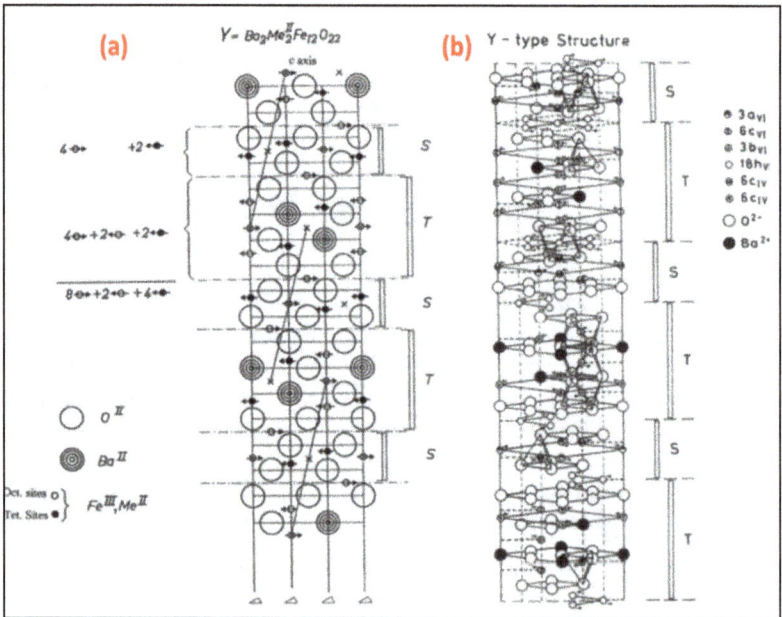

Figure 9: (a) Cross-sectional (b) Perspective view of structure of Y-type hexaferrite(Copyright permission with the license number: 5146980447011) [21]

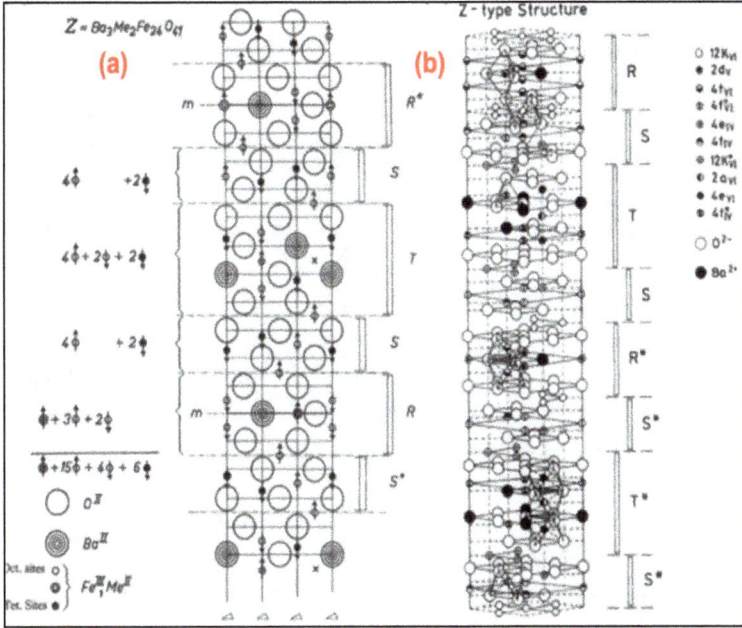

Figure 10: (a) Cross-sectional (b) Perspective view of structure of Z-type Hexaferrite (Copyright permission with the license number: 5146980447011) [21]

According to the periodic table, a Z-type hexaferrite unit cell has 140 atoms and is placed in the $P6_3/mmc$ space group. Numerous metal ions, including Fe^{3+}, Co^{2+}, Zn^{2+}, and Cu^{2+}, occupy non-equivalent interstitial spaces. Z-type hexaferrite has tremendous promise as an anti-EMI material or as a radar absorbent material (RAM) for high-frequency magnetic devices. Due to their ferromagnetic resonance at high GHz frequencies, these ferrites are employed in inductor cores for UHF communications in microwaves [82,83].

5.6   U-type hexaferrite

The U-type hexaferrite, chemical formula $Ba_4Me_2Fe_{36}O_{60}$ (Me2U, where 'Me' is any of the following elements: Cu, Fe, Co, Mn, Mg, and so on), has the most intricate crystal structure and largest unit cell size of all ferrites. Due to the material's complex crystal structure, single-phase production of U-type hexaferrite is challenging. As seen in the image, their crystal structure is dictated by the stacking sequence RSR*S*TS* with the space group R3 m. The rhombohedral structure of the unit cell of a three-molecule U-type chemical is identified in the space group R3m. Two M-blocks and one Y-block are stacked along the c-axis to make the structure. U-type hexaferrite has exceptional electromagnetic characteristics in the microwave region [22], and this is especially true in the microwave range. For millimeter-wave applications, U-type hexaferrite is a material worth exploring [81, 84–86]. The cross-sectional picture of U-type hexaferrite is shown in Figure 11.

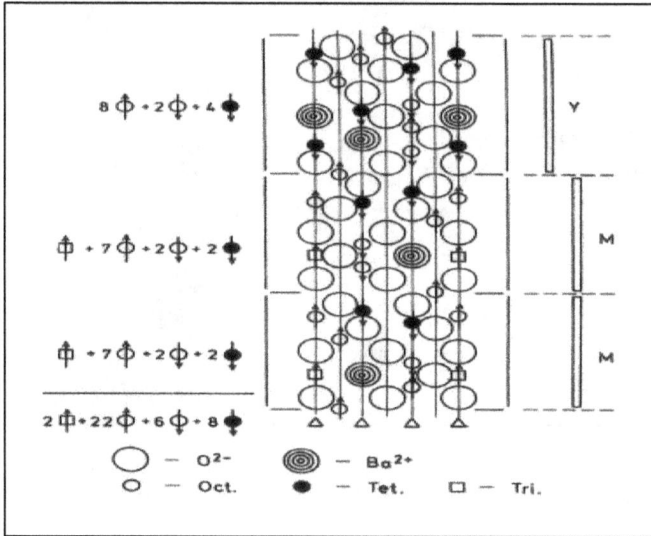

Figure 11: Cross-sectional view of structure of U-type Hexaferrite(Copyright permission with the license number: 5146980447011) [21]

6.   Applications of hexaferrites

The magnetic materials are utilized in various applications, for instance, motors, transformers, annetas, defence system, diagnostic devices, communication systems, storage devices, recording media, electron beams concentration, sensors, etc. [28,87,88]. The ferromagnetic as well as ferrimagnetic ceramic based materials are the widely used class of magnetic materials for this purpose [7]. Out of all these ceramic based materials, hexaferrites are utilized on a large scale and therefore, few of their applications are presented in the schematic representation as provided in figure 12.

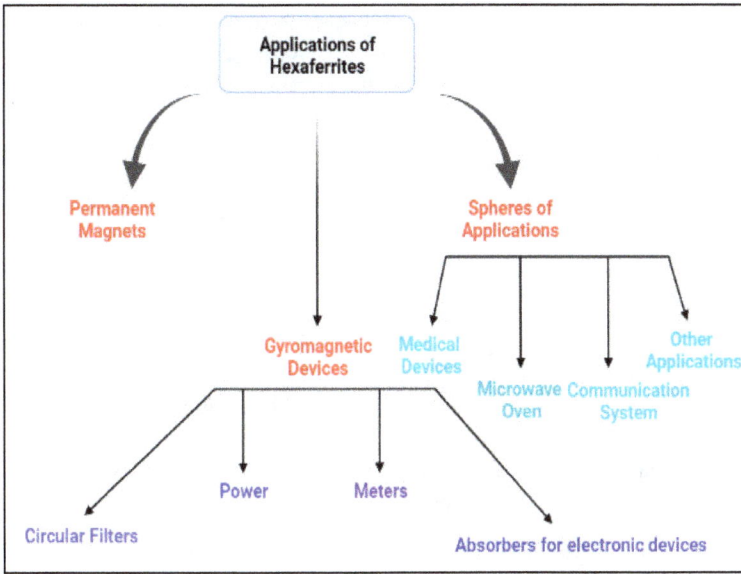

Figure 12: Applications of hexaferrites

Concluding Remarks

In the current chapter, we have reviewed the historical progresses of hexaferrites along with their classification, structure and applications in various research areas. Two approaches called as spinel based and S/R/T blocks-based model have been taken into practice for studying the structure of hexagonal ferrites. Under the crystallographic structure, each hexaferrite is made of various S/R/T blocks having different space geometry. The M-type/Z-type/W-type is made of RSR*S*/STSR/RSSR*S*S* blocks model having P63/mmc space geometry whereas the X-type/Y-type/U-type hexaferrite consists of SRS*S*R*/ST/RSR*S*TS* blocks with R$\bar{3}$m geometry respectively. Lastly, the applications of hexagonal ferrites in different research areas are provided.

References

[1]   R. Jasrotia, V.P. Singh, R.K. Sharma, M. Singh, Analysis of optical and magnetic study of silver substituted SrW hexagonal ferrites, in: AIP Conference Proceedings, AIP Publishing LLC, 2019: p. 090004. https://doi.org/10.1063/1.5122448

[2]   V.P. Singh, R. Jasrotia, R. Kumar, P. Raizada, S. Thakur, K.M. Batoo, M. Singh, A current review on the synthesis and magnetic properties of M-type hexaferrites material, World Journal of Condensed Matter Physics. 8 (2018) 36. https://doi.org/10.4236/wjcmp.2018.82004

[3]   R. Jasrotia, P. Puri, V.P. Singh, R. Kumar, Sol–gel synthesized Mg–Ag–Mn nanoferrites for power applications, Journal of Sol-Gel Science and Technology. 97 (2021) 205–212. https://doi.org/10.1007/s10971-020-05428-3

[4]   S. Kour, R.K. Sharma, R. Jasrotia, V.P. Singh, A brief review on the synthesis of maghemite (γ-Fe2O3) for medical diagnostic and solar energy applications, in: AIP Conference Proceedings, AIP Publishing LLC, 2019: p. 090007. https://doi.org/10.1063/1.5122451

[5]   S.V. Bhandare, R. Kumar, A.V. Anupama, M. Mishra, R.V. Kumar, V.M. Jali, B. Sahoo, Effect of Mg-substitution in Co–Ni-Ferrites: Cation distribution and magnetic properties, Materials Chemistry and Physics. 251 (2020) 123081. https://doi.org/10.1016/j.matchemphys.2020.123081

[6]   R. Jasrotia, V.P. Singh, R. Kumar, R. Verma, A. Chauhan, Effect of Y3+, Sm3+ and Dy3+ ions on the microstructure, morphology, optical and magnetic properties NiCoZn magnetic nanoparticles, Results in Physics. 15 (2019) 102544. https://doi.org/10.1016/j.rinp.2019.102544

[7]   K. Dulta, G.K. Ağçeli, P. Chauhan, R. Jasrotia, P.K. Chauhan, A novel approach of synthesis zinc oxide nanoparticles by bergenia ciliata rhizome extract: antibacterial and anticancer potential, Journal of Inorganic and Organometallic Polymers and Materials. 31 (2021) 180–190. https://doi.org/10.1007/s10904-020-01684-6

[8]   R. Jasrotia, G. Kumar, K.M. Batoo, S.F. Adil, M. Khan, R. Sharma, A. Kumar, V.P. Singh, Synthesis and characterization of Mg-Ag-Mn nano-ferrites for electromagnet applications, Physica B: Condensed Matter. 569 (2019) 1–7. https://doi.org/10.1016/j.physb.2019.05.033

[9]   R. Jasrotia, J. Prakash, G. Kumar, R. Verma, S. Kumari, S. Kumar, V.P. Singh, A.K. Nadda, S. Kalia, Robust and sustainable Mg1-xCexNiyFe2-yO4 magnetic nanophotocatalysts with improved photocatalytic performance towards photodegradation of crystal violet and rhodamine B pollutants, Chemosphere. (2022) 133706. https://doi.org/10.1016/j.chemosphere.2022.133706

[10]   S. Kour, R. Jasrotia, P. Puri, A. Verma, B. Sharma, V.P. Singh, R. Kumar, S. Kalia, Improving photocatalytic efficiency of MnFe 2 O 4 ferrites via doping with Zn 2+/La 3+ ions: photocatalytic dye degradation for water remediation, Environmental

Science and Pollution Research. (2021) 1–16. https://doi.org/10.1007/s11356-021-13147-7

[11]   R. Jasrotia, Spinel Ferrite Based Nanomaterials for Water Remediation Application, Materials Research Foundations. 112 (n.d.).

[12]   J.J. Went, Ferroxdure. a class of new permanent magnet materials, Phil. Tech. Rev. 13 (1952) 361.

[13]   P. Dhiman, Hexagonal Ferrites, Synthesis, Properties and Their Applications, Materials Research Foundations. 112 (n.d.).

[14]   R. Jasrotia, P. Puri, A. Verma, V.P. Singh, Magnetic and electrical traits of sol-gel synthesized Ni-Cu-Zn nanosized spinel ferrites for multi-layer chip inductors application, Journal of Solid State Chemistry. 289 (2020) 121462. https://doi.org/10.1016/j.jssc.2020.121462

[15]   M.T. Weiss, P.W. Anderson, Ferromagnetic resonance in ferroxdure, Physical Review. 98 (1955) 925. https://doi.org/10.1103/PhysRev.98.925

[16]   K.J. Sixtus, K.J. Kronenberg, R.K. Tenzer, Investigations on barium ferrite magnets, Journal of Applied Physics. 27 (1956) 1051–1057. https://doi.org/10.1063/1.1722540

[17]   V.G. Harris, Z. Chen, Y. Chen, S. Yoon, T. Sakai, A. Gieler, A. Yang, Y. He, K.S. Ziemer, N.X. Sun, Ba-hexaferrite films for next generation microwave devices, Journal of Applied Physics. 99 (2006) 08M911. https://doi.org/10.1063/1.2165145

[18]   R. Jasrotia, V.P. Singh, R. Kumar, M. Singh, Raman spectra of sol-gel auto-combustion synthesized Mg-Ag-Mn and Ba-Nd-Cd-In ferrite based nanomaterials, Ceramics International. 46 (2020) 618–621. https://doi.org/10.1016/j.ceramint.2019.09.012

[19]   M. Chandel, V.P. Singh, R. Jasrotia, K. Singha, R. Kumar, A review on structural, electrical and magnetic properties of Y-type hexaferrites synthesized by different techniques for antenna applications and microwave absorbing characteristic materials, AIMS Materials Science. 7 (2020) 244–268. https://doi.org/10.3934/matersci.2020.3.244

[20]   S.K. Godara, V. Kaur, P.S. Malhi, M. Singh, S. Verma, R. Jasrotia, J. Ahmed, M.S. Tamboli, A.K. Sood, Investigation of microstructural and magnetic properties of Ca2+ doped strontium hexaferrite nanoparticles, Journal of King Saud University-Science. (2022) 101963. https://doi.org/10.1016/j.jksus.2022.101963

[21]   R.C. Pullar, Hexagonal ferrites: a review of the synthesis, properties and applications of hexaferrite ceramics, Progress in Materials Science. 57 (2012) 1191–1334. https://doi.org/10.1016/j.pmatsci.2012.04.001

[22]   R. Jasrotia, R. Khargotra, A. Verma, I. Sharma, R. Verma, Applications of Multiferroics, in: Ferrites and Multiferroics, Springer, 2021: pp. 195–213. https://doi.org/10.1007/978-981-16-7454-9_12

[23]   V. Adelskold, Arkiv Kemi Miner, (1938).

[24]   J.J. Went, G.W. Rathenau, E.W. Gorter, van Oosterhout, GW, Philips Tech. Rev. 13. 194 (1952).

[25]   J. Smit, H.P.J. Wijn, Ferrites, Philips technical library, Eindhoven, The Netherlands. 278 (1959).

[26]   R. Jasrotia, V.P. Singh, R. Kumar, K. Singha, M. Chandel, M. Singh, Analysis of Cd^{2+} and In^{3+} ions doping on microstructure, optical, magnetic and mo\" ssbauer spectral properties of sol-gel synthesized BaM hexagonal ferrite based nanomaterials, Results in Physics. 12 (2019) 1933–1941. https://doi.org/10.1016/j.rinp.2019.01.088

[27]   R. Verma, A. Chauhan, K.M. Batoo, R. Jasrotia, A. Sharma, R. Kumar, M. Hadi, E.H. Raslan, J.P. Labis, A. Imran, Modulation of Dielectric, Ferroelectric, and Piezoelectric Properties of Lead-Free BCZT Ceramics by Doping, ECS Journal of Solid State Science and Technology. (2021). https://doi.org/10.1149/2162-8777/ac0e0d

[28]   R. Jasrotia, V.P. Singh, B. Sharma, A. Verma, P. Puri, R. Sharma, M. Singh, Sol-gel synthesized Ba-Nd-Cd-In nanohexaferrites for high frequency and microwave devices applications, Journal of Alloys and Compounds. 830 (2020) 154687. https://doi.org/10.1016/j.jallcom.2020.154687

[29]   P.D. File, S. Kabekkodu, International Centre for Diffraction Data: Newtown Square, PA, USA. (2004).

[30]   Z. Li, F. Gao, Chemical bond and hardness of M-, W-type hexagonal barium ferrites, Canadian Journal of Chemistry. 89 (2011) 573–576. https://doi.org/10.1139/v11-013

[31]   K. Singha, R. Jasrotia, V.P. Singh, M. Chandel, R. Kumar, S. Kalia, A study of magnetic properties of Y–Ni–Mn substituted Co 2 Z-type nanohexaferrites via vibrating sample magnetometry, Journal of Sol-Gel Science and Technology. 97 (2021) 373–381. https://doi.org/10.1007/s10971-020-05412-x

[32] G. Albanese, M. Carbucicchio, A. Deriu, G. Asti, S. Rinaldi, Influence of the cation distribution on the magnetization of Y-type hexagonal ferrites, Applied Physics. 7 (1975) 227–238. https://doi.org/10.1007/BF00936028

[33] M. Chandel, V.P. Singh, R. Jasrotia, K. Singha, M. Singh, P. Thakur, S. Kalia, Fabrication of Ni2+ and Dy3+ substituted Y-Type nanohexaferrites: a study of structural and magnetic properties, Physica B: Condensed Matter. 595 (2020) 412378. https://doi.org/10.1016/j.physb.2020.412378

[34] L.M. Castelliz, K.M. Kim, P.S. Boucher, Preparation, stability range and high frequency permeability of some ferroxplana compounds, J. Can. Ceram. Soc. 38 (1969) 57.

[35] R. Jasrotia, V.P. Singh, R.K. Sharma, P. Kumar, M. Singh, Analysis of effect of Ag+ ion on microstructure and elemental distribution of strontium W-type hexaferrites, in: AIP Conference Proceedings, AIP Publishing LLC, 2019: p. 140004. https://doi.org/10.1063/1.5122517

[36] Ü. Özgür, Y. Alivov, H. Morkoç, Microwave ferrites, part 1: fundamental properties, Journal of Materials Science: Materials in Electronics. 20 (2009) 789–834. https://doi.org/10.1007/s10854-009-9923-2

[37] A. Cochardt, Modified strontium ferrite, a new permanent magnet material, Journal of Applied Physics. 34 (1963) 1273–1274. https://doi.org/10.1063/1.1729468

[38] H.P.J. Wijn, A new method of melting ferromagnetic semiconductors. BaFe 18 O 27, a new kind of ferromagnetic crystal with high crystal anisotropy, Nature. 170 (1952) 707–708. https://doi.org/10.1038/170707a0

[39] P.B. Braun, Crystal structure of BaFe18O27, Nature. 170 (1952) 708–708. https://doi.org/10.1038/170708a0

[40] N. Tran, D.H. Kim, B.W. Lee, Influence of Fabrication Conditions on the Structural and the Magnetic Properties of Co-doped BaFe 12 O 19 Hexaferrites, Journal of the Korean Physical Society. 72 (2018) 731–736. https://doi.org/10.3938/jkps.72.731

[41] R. Valenzuela, Novel applications of ferrites, Physics Research International. 2012 (2012). https://doi.org/10.1155/2012/591839

[42] G. Albanese, Recent advances in hexagonal ferrites by the use of nuclear spectroscopic methods, Le Journal de Physique Colloques. 38 (1977) C1-85. https://doi.org/10.1051/jphyscol:1977117

[43]   D. Lisjak, P. McGuiness, M. Drofenik, Thermal instability of Co-substituted barium hexaferrites with U-type structure, Journal of Materials Research. 21 (2006) 420–427. https://doi.org/10.1557/jmr.2006.0048

[44]   L. Alhmoud, A.R. Al Dairy, H. Faris, I. Aljarah, Prediction of Hysteresis Loop of Barium Hexaferrite Nanoparticles Based on Neuroevolutionary Models, Symmetry. 13 (2021) 1079. https://doi.org/10.3390/sym13061079

[45]   H. Belrhazi, M.Y.E. Hafidi, M.E. Hafidi, Permanent magnets elaboration from BaFe 12 O 19 hexaferrite material: Simulation and prototype, Res Dev Mater Sci. 11 (2019) 1–5. https://doi.org/10.31031/RDMS.2019.11.000757

[46]   G. Muleta, The study of optical, electrical and dielectric properties of cadmium and zinc substituted copper ferrite nanoparticles, Ethiopia: Arba Minch University. (2018).

[47]   S. Arcaro, Modern Ferrites in Engineering: Synthesis, Processing and Cutting-Edge Applications, Springer Nature, n.d.

[48]   A. Goldman, Modern ferrite technology, Springer Science & Business Media, 2006.

[49]   G.F. Dionne, Magnetic Ions in Oxides, in: Magnetic Oxides, Springer, 2009: pp. 37–106. https://doi.org/10.1007/978-1-4419-0054-8_2

[50]   S. Rösler, P. Wartewig, H. Langbein, Synthesis and characterization of hexagonal ferrites BaFe12-2xZnxTixO19 (0≤ x≤ 2) by thermal decomposition of freeze-dried precursors, Crystal Research and Technology: Journal of Experimental and Industrial Crystallography. 38 (2003) 927–934. https://doi.org/10.1002/crat.200310115

[51]   M.W. Pieper, A. Morel, F. Kools, NMR analysis of La+ Co doped M-type ferrites, Journal of Magnetism and Magnetic Materials. 242 (2002) 1408–1410. https://doi.org/10.1016/S0304-8853(01)00963-5

[52]   A. Morel, J.M. Le Breton, J. Kreisel, G. Wiesinger, F. Kools, P. Tenaud, Sublattice occupation in Sr1- xLaxFe12- xCoxO19 hexagonal ferrite analyzed by Mössbauer spectrometry and Raman spectroscopy, Journal of Magnetism and Magnetic Materials. 242 (2002) 1405–1407. https://doi.org/10.1016/S0304-8853(01)00962-3

[53]   R.B. Jotania, H.S. Virk, Y-type Hexaferrites: structural, dielectric and magnetic properties, in: Solid State Phenomena, Trans Tech Publ, 2012: pp. 209–232. https://doi.org/10.4028/www.scientific.net/SSP.189.209

[54]   L. Lechevallier, J.M. Le Breton, J.F. Wang, I.R. Harris, Structural analysis of hydrothermally synthesized Sr1- xSmxFe12O19 hexagonal ferrites, Journal of Magnetism and Magnetic Materials. 269 (2004) 192–196. https://doi.org/10.1016/S0304-8853(03)00591-2

[55]   M. Pardavi-Horvath, Microwave applications of soft ferrites, Journal of Magnetism and Magnetic Materials. 215 (2000) 171–183. https://doi.org/10.1016/S0304-8853(00)00106-2

[56]   V.G. Harris, A. Geiler, Y. Chen, S.D. Yoon, M. Wu, A. Yang, Z. Chen, P. He, P.V. Parimi, X. Zuo, Recent advances in processing and applications of microwave ferrites, Journal of Magnetism and Magnetic Materials. 321 (2009) 2035–2047. https://doi.org/10.1016/j.jmmm.2009.01.004

[57]   F.K. Lotgering, P. Vromans, M.A.H. Huyberts, Permanent-magnet material obtained by sintering the hexagonal ferrite W= BaFe18O27, Journal of Applied Physics. 51 (1980) 5913–5918. https://doi.org/10.1063/1.327493

[58]   A. Paoluzi, F. Licci, O. Moze, G. Turilli, A. Deriu, G. Albanese, E. Calabrese, Magnetic, Mössbauer, and neutron diffraction investigations of W-type hexaferrite BaZn2- x Co x Fe16O27 single crystals, Journal of Applied Physics. 63 (1988) 5074–5080. https://doi.org/10.1063/1.340405

[59]   E.P. Wohlfarth, A.S. Arrott, Ferromagnetic materials: a handbook on the properties of magnetically ordered substances, Vols. 1 and 2, Physics Today. 35 (1982) 63. https://doi.org/10.1063/1.2914974

[60]   M.A. Ahmed, N. Okasha, R.M. Kershi, Extraordinary role of rare-earth elements on the transport properties of barium W-type hexaferrite, Materials Chemistry and Physics. 113 (2009) 196–201. https://doi.org/10.1016/j.matchemphys.2008.07.032

[61]   S. Ruan, B. Xu, H. Suo, F. Wu, S. Xiang, M. Zhao, Microwave absorptive behavior of ZnCo-substituted W-type Ba hexaferrite nanocrystalline composite material, Journal of Magnetism and Magnetic Materials. 212 (2000) 175–177. https://doi.org/10.1016/S0304-8853(99)00755-6

[62]   D. Samaras, A. Collomb, S. Hadjivasiliou, C. Achilleos, J. Tsoukalas, J. Pannetier, J. Rodriguez, The rotation of the magnetization in the BaCo2Fe16O27 W-type hexagonal ferrite, Journal of Magnetism and Magnetic Materials. 79 (1989) 193–201. https://doi.org/10.1016/0304-8853(89)90098-X

[63] E.P. Naiden, G.I. Ryabtsev, Magnetization processes of the first kind in the hexaferrite Co 0.62 Zn 1.38 W, Soviet Physics Journal. 33 (1990) 318–321. https://doi.org/10.1007/BF00894211

[64] C. Sürig, K.A. Hempel, R. Müller, P. Görnert, Investigations on Zn2- xCoxW-type hexaferrite powders at low temperatures by ferromagnetic resonance, Journal of Magnetism and Magnetic Materials. 150 (1995) 270–276. https://doi.org/10.1016/0304-8853(95)00406-8

[65] E.P. Naiden, S.M. Zhilyakov, Investigation of the magnetocaloric effect in hexagonal ferrimagnetic materials with spin-orientational transitions, Russian Physics Journal. 40 (1997) 869–874. https://doi.org/10.1007/BF02523101

[66] G. Albanese, M. Carbucicchio, G. Asti, Spin-order and magnetic properties of BaZn 2 Fe 16 O 27 (Zn 2-W) hexagonal ferrite, Applied Physics. 11 (1976) 81–88. https://doi.org/10.1007/BF00895020

[67] A. Lilot, A. Gérard, F. Grandjean, Analysis of the superexchange interactions paths in the W-hexagonal ferrites, IEEE Transactions on Magnetics. 18 (1982) 1463–1465. https://doi.org/10.1109/TMAG.1982.1062044

[68] A. Collomb, O. Abdelkader, P. Wolfers, J.C. Guitel, D. Samaras, Crystal structure and magnesium location in the W-type hexagonal ferrite:[Ba] Mg2-W, Journal of Magnetism and Magnetic Materials. 58 (1986) 247–253. https://doi.org/10.1016/0304-8853(86)90444-0

[69] L. De-Xin, Z. Nan-Nin, G. Shu-Jiao, L. Guo-Dong, W. Hui-Zong, Magnetic and Mossbauer study of (TiCu) Ni/sub 2/W hexagonal ferrite system, IEEE Transactions on Magnetics. 25 (1989) 3290–3292. https://doi.org/10.1109/20.42281

[70] E.J.W. Verwey, P.W. Haaijman, F.C. Romeijn, G.W. Vanoosterhout, Controlled-valency semiconductors, Philips Research Reports. 5 (1950) 173–187.

[71] P.B. Braun, The crystal structures of a new group of ferromagnetic compounds, Philips Res. Rep. 12 (1957) 491–548.

[72] F. Leccabue, R. Panizzieri, G. Bocelli, G. Calestani, C. Rizzoli, N.S. Almodovar, Crystal structure and magnetic characterization of Sr2Zn2Fe28O46 (SrZn- X) hexaferrite single crystal, Journal of Magnetism and Magnetic Materials. 68 (1987) 365–373. https://doi.org/10.1016/0304-8853(87)90015-1

[73] Z. Haijun, Y. Xi, Z. Liangying, The preparation and microwave properties of Ba2ZnxCo2- xFe28O46 hexaferrites, Journal of Magnetism and Magnetic Materials. 241 (2002) 441–446. https://doi.org/10.1016/S0304-8853(01)00447-4

[74]   S. Ishiwata, Y. Taguchi, H. Murakawa, Y. Onose, Y. Tokura, Low-magnetic-field control of electric polarization vector in a helimagnet, Science. 319 (2008) 1643–1646. https://doi.org/10.1126/science.1154507

[75]   T. Kimura, G. Lawes, A.P. Ramirez, Electric polarization rotation in a hexaferrite with long-wavelength magnetic structures, Physical Review Letters. 94 (2005) 137201. https://doi.org/10.1103/PhysRevLett.94.137201

[76]   Y. Bai, J. Zhou, Z. Gui, L. Li, Phase formation process, microstructure and magnetic properties of Y-type hexagonal ferrite prepared by citrate sol–gel auto-combustion method, Materials Chemistry and Physics. 98 (2006) 66–70. https://doi.org/10.1016/j.matchemphys.2005.08.067

[77]   M. Obol, X. Zuo, C. Vittoria, Oriented Y-type hexaferrites for ferrite device, Journal of Applied Physics. 91 (2002) 7616–7618. https://doi.org/10.1063/1.1446113

[78]   T. Nakamura, K. Hatakeyama, Complex permeability of polycrystalline hexagonal ferrites, IEEE Transactions on Magnetics. 36 (2000) 3415–3417. https://doi.org/10.1109/20.908844

[79]   H.J. Kwon, J.Y. Shin, J.H. Oh, The microwave absorbing and resonance phenomena of Y-type hexagonal ferrite microwave absorbers, Journal of Applied Physics. 75 (1994) 6109–6111. https://doi.org/10.1063/1.355476

[80]   M. Obol, C. Vittoria, Microwave permeability of Y-type hexaferrites in zero field, Journal of Applied Physics. 94 (2003) 4013–4017. https://doi.org/10.1063/1.1601291

[81]   H.A. Elkady, M.M. Abou-Sekkina, K. Nagorny, New information on Mössbauer and phase transition properties of Z-type hexaferrites, Hyperfine Interactions. 128 (2000) 423–432. https://doi.org/10.1023/A:1012612405813

[82]   H. Zhang, J. Zhou, Z. Yue, P. Wu, Z. Gui, L. Li, Synthesis of Co2Z hexagonal ferrite with planar structure by gel self-propagating method, Materials Letters. 43 (2000) 62–65. https://doi.org/10.1016/S0167-577X(99)00231-1

[83]   X. Wang, L. Li, S. Su, Z. Gui, Z. Yue, J. Zhou, Low-temperature sintering and high frequency properties of Cu-modified Co2Z hexaferrite, Journal of the European Ceramic Society. 23 (2003) 715–720. https://doi.org/10.1016/S0955-2219(02)00157-7

[84]   A.J. Kerecman, A. Tauber, T.R. AuCoin, R.O. Savage, Magnetic properties of Ba 4 Zn 2 Fe 36 O 60 single crystals, Journal of Applied Physics. 39 (1968) 726–727. https://doi.org/10.1063/1.2163602

[85]   A.J. Kerecman, T.R. AuCoin, W.P. Dattilo, Ferromagnetic Resonance in Ba4Zn2Fe36O60 (ZnU) and Mn-Substituted ZnU Single Crystals, Journal of Applied Physics. 40 (1969) 1416–1417. https://doi.org/10.1063/1.1657698

[86]   G. Albanese, A. Deriu, F. Licci, S. Rinaldi, Preparation and magnetic characterization of the Ba 2 Zn 2-2x Cu 2x Fe 12 O 22 hexagonal ferrites, IEEE Transactions on Magnetics. 14 (1978) 710–712. https://doi.org/10.1109/TMAG.1978.1059812

[87]   R. Jasrotia, S. Kour, P. Puri, A.D. Jara, B. Singh, C. Bhardwaj, V.P. Singh, R. Kumar, Structural and magnetic investigation of Al3+ and Cr3+ substituted Ni–Co–Cu nanoferrites for potential applications, Solid State Sciences. 110 (2020) 106445. https://doi.org/10.1016/j.solidstatesciences.2020.106445

[88]   R. Jasrotia, N. Kumari, R. Kumar, M. Naushad, P. Dhiman, G. Sharma, Photocatalytic degradation of environmental pollutant using nickel and cerium ions substituted Co 0.6 Zn 0.4 Fe 2 O 4 nanoferrites, Earth Systems and Environment. (2021) 1–19. https://doi.org/10.1007/s41748-021-00214-9

An Introduction to Hard Ferrites: From Fundamentals to Practical Applications   Materials Research Forum LLC
Materials Research Foundations **142** (2023) 35-65   https://doi.org/10.21741/9781644902318-2

Chapter 2

Recent Advances in Processing of Hard Ferrites

Garima Rana[1*], Pooja Dhiman[1], Rahul Kalia[2], Ankush Chauhan[3], RiteshVerma[2], Rajesh Kumar[2]

[1]International Research Centre of Nanotechnology for Himalayan Sustainability (IRCNHS), Shoolini University, India

[2]School of Physics & Materials Science, Shoolini University of Biotechnology and Management Sciences, Bajhol, Solan (H.P.)

[3]Faculty of Allied Health Sciences, Chettinad Academy of Research and Education, Kelambakkam, Changalpattu, Tamil Nadu, India-60310

*mrs.garimarana@gmail.com

Abstract

We plan to discuss possible fabrication approaches for hard ferrites, which are well-known for their vast range of applications and uses. We've covered all six varieties of hexagonal ferrites in this chapter: M-type, Z-type, Y-type, W-type, X-type, and U-type hexa-ferrites. Hexaferrites are ferromagnetic materials, and their properties are entirely determined by the intrinsic structure of ferrites. The numerous synthesis procedures for hard ferrite nanoparticles are the focus of this chapter. This chapter describes in detail the different processes for preparing hard ferrites, including examples, advantages, and limitations. Solid-state, combustion, hydrothermal, microemulsion, solvothermal, solution combustion, biosynthetic approach, microwave-assisted combustion, sol–gel, co–precipitation, or laser pyrolysis, sonochemical, thermal decomposition, and reverse micelle processes are briefly explained. The influence of synthesis procedures on the characteristics of ferrites in nanoform is summarised in this chapter.

Keywords

Hard Ferrites, Nanomaterials, Synthesis Methods

Contents

1. Introduction

In terms of applications, ferrite is a particularly important category of magnetic particle. In 1930, ferrite was first used in practical applications, and its structural, electrical, and magnetic properties were studied. Many researchers have researched ferrite extensively [1]. In 1951, the world's first permanent magnet, based on ferroxdure - $BaFe_{12}O_{19}$ (commonly known as BaM) [2], was created. Scientists and technologists have demonstrated a strong interest in investigating hexagonal ferrites since their discovery in the 1950s, and this interest continues today. In 1955, the application of gyromagnetic properties of hexaferrite was systematically begun [3-5]. Recently, there is a lot of work done in nanotechnology magneto electric/multiferroic applications and the development of orientation and alignment effect in ferrite in fibers [6]. Hexagonal ferrites have a wide range of applications, both commercially and technically. Also included in these materials is the majority of the total magnetic material, which is widely used and manufactured around the world [7]. The material's extremely high specific resistance and extraordinary magnetic flexibility make it an excellent candidate for ferrite operational frequencies in telecommunications, electronic devices, and microwave/GHz frequencies [8]. Ferrites are used as permanent magnets in the market due to their low price, high magnetic performance, and have attracted attention over the years. Ferrites are very sensitive to the research method, sintering condition, amount of components metal oxide, various additive dopants and impurities include [9, 10]. Hexaferrites are ferrites with high saturation magnetization, coercivity, and remanent magnetization values. Ferrite is a magnetoplambite structure, also known as hexagonal structure, with the chemical formula $AFe_{12}O_{19}$, where 'A' is the Ba and Sr. The regular high ceramic methods are used to make ferrites, however extensive heating regimens at high temperatures limit the creation of small particles and reduce homogeneity [11]. Wet chemical methods, which provide the benefits of the creation of tiny particles with tailored chemical compositions, high purity, greater homogeneity, and controlled shape and structure, are gradually supplanting this way. The microstructure-dependent properties, especially electrical and magnetic properties, are affected by good management of these characteristics. Controlling the size and surface area of materials requires the use of synthesis processes. The properties of all ferrites depend not only on their chemical composition but also on the methods used for the preparation. However, variations in the different properties can be made by adding a little bit of impurity and correspondingly their use to cover several applications. This hexagonal ferrite feature increases its adaptability to a wide range of applications. This ferrite forms a hexagonal configuration when it crystallizes. Apart from spinel ferrites, these ferrites' magnetic structure allows the trait to operate across the whole gigahertz band because of their substantial essential magnetic variance[12].M, Z, Y, W, X, and U are the six different varieties of hexagonal ferrite.

M-type hexaferrite

M-type ferrites were the first hexaferrite to be discovered, and they remain the most economically relevant material [13]. It was first studied and magnetically characterized in the early 1950s by Phillips [14]. M-type hexaferrite is a solid solution with the same structure as magnetoplumbite and is written in the molecular form $M_eO.6Fe_2O_3$ or $MeFe_{12}O_{19}$. Where, Me can be the divalent ions Ba^{2+}, Sr^{2+} or Pb^{2+} [15-17]. The magnetoplumbite structure is made up of spinel blocks with two oxygen layers, S and S*, linked by a block R containing barium or strontium ion. The crystallographic structure can be described as RSR*S* and the space group is denoted as P6₃/mmc [16]. The hexagonal structure of M-type hexaferrite crystallizes with 64 ions per unit cell on 11 different symmetry sites. Therefore 38 oxygen ions, 24 ferric ions, and 2 Me ions (Me = Ba^{2+}, Sr^{2+}, Pb^{2+}, and La^{3+}) are present in the unit cell.

W-type hexagonal ferrite

In 1980, Philip's research laboratories discovered the Me_2W (W-type) $BaMe_2Fe_{16}O_{27}$ hexaferrite [18]. It exhibits a magnetic uniaxial anisotropy along the axis of the hexagonal structure as in the case of the M-type hexaferrite. More and more attention has been devoted to this hexaferrite in the past decades because of its moderate magnetic properties, excellent chemical stability, low cost, high coercivity, and large magnetic energy product [19, 20]. The W-type structure is closely related to the M structure because it has one Ba-containing layer for every seventh oxygen layer in the spinel structure. The saturation magnetization and Néel temperature of W hexaferrite may be changed by substituting different combinations of divalent metal ions. The crystal structure of W-type hexagonal ferrite is very complex and can be considered as a superposition of R and S blocks (where R = $BaFe_6O_{11}^{2-}$ is hexagonal and S =$Fe_6O_8^{2+}$ is cubic spinel) along the hexagonal c-axis with a structure of RSSR*S*S*, where R is three oxygen layer block with composition $BaFe_6O_{11}$, S (Spinel block) is a two oxygen layer block with composition Fe_6O_8 and '*' means the respective block is turned 180° around the hexagonal axis. The structural and magnetic properties of W-type hexaferrite depend on many factors like the method of preparation, sintering temperature, type and amount of substitution, etc. [21, 22].

X-type hexaferrite

The chemical formula for X-type hexaferrite is $Ba_2Me_2Fe_{28}O_{46}$, and it was discovered almost 50 years ago. Braun published the crystal structure of $Ba_2Fe_{30}O_{46}$, which included real-space coordinates for oxygen, barium, and transition metal ions [23]. X-type hexagonal ferrites have a structure that can be described as a stacking of R- and S-blocks along the hexagonal c-axis with a model of RSRSSR*S*R*S*S*, where R is a three-oxygen layer block with composition $BaFe_6O_{11}$, S (spinel block) is a two-oxygen-layer block with composition Fe_6O_8, and the asterisk indicates that the corresponding block has been turned 180° around the hexagonal axis. X-type phases are usually seen mixed with M-type and W-type phases and are difficult to separate. Three chemical formula units (Z = 3) make up the elementary cell (a = 5.88, c = 84.11, density = 5.30 g/cm³). The unit cell is made up of four different layers of M and W structures belonging to the R$\bar{3}$m space symmetry [24, 25].

Y-type hexagonal ferrite

The general formula of Y-type hexaferrite is $Ba_2Me_2Fe_{12}O_{22}$. The crystalline structure of these ferrites is made up of blocks with a spinel structure and blocks containing Ba-O laminated alternatively on top of each other. Because the space group is a rhombohedral crystal R$\bar{3}$m, each spinel layer, and Ba-O block is repeated three times in a hexagonal practical unit cell. The average valence of the Fe ion is three, with oxygen positioned at the vertices of an octahedron or a tetrahedron since Ba represents alkaline earth metals and Me represents divalent ions like Mg or Zn. These two and four layers blocks are termed as S and T respectively. Further stacking gives the ST unit. The unit cell with hexagonal symmetry consists of 18 oxygen layers with a repeat distance extending through only six oxygen layers, the length of the c- axis being 43.56 Å. In the hexagonal element cell, each layer again contains four large ions. There are four successive layers of four oxygen ions, followed by two layers each containing three oxygen ions and one barium ion. The unit cell is composed of the sequence STSTST including three formula units.

Z-type hexaferrite

Z-type hexaferrite is one of the most complicated compounds in the hexaferrite family, with high permeability up to GHz, high resistivity, and high thermal and chemical stability. M- ($BaFe_{12}O_{19}$) and Y- ($Ba_2Me_2Fe_{12}O_{22}$) type hexaferrite is combined to form Z-hexaferrite. The crystal structure is made up of 33 layers that are stacked along the hexagonal c-axis and organized into S, R, and T blocks [26]. Z-type hexaferrite unit cell has 140 atoms and belongs to the P6$_3$/mmc space group. Non-equivalent interstitial spaces are occupied by metal ions such as Fe^{3+}, Co^{2+}, Zn^{2+}, and Cu^{2+}. Z-type hexaferrite has great potential applications as anti-EMI material, radar absorbing materials (RAM) for magnetic devices in the GHz region. These ferrites are very useful for inductor cores or in UHF communications in the microwave region because of ferromagnetic resonance at GHz frequencies [27, 28].

U-type hexaferrite

The most complicated crystal structure and greatest unit cell size are found in U-type hexaferrite, $Ba_4Me_2Fe_{36}O_{60}$ (Me_2U, Me = Cu, Fe, Co, Mn, Mg, etc.). Due to their complicated crystal structure, U-type hexaferrite is difficult to produce in a single phase. They are a superposition of M and Y-blocks along the c-axis in terms of composition and structure, and their crystal structure is represented by the stacking sequence RSR*S*TS* with the space group R$\bar{3}$m. The rhombohedral structure of the unit cell of the U-type compound produced by three molecules belongs to the space group R$\bar{3}$m. Along the c-axis, two M-blocks and one Y-block are superimposed to create the structure. In the microwave range, U-type hexaferrite has outstanding electromagnetic properties [29].

2.   Fabrication of hard ferrites nanoparticles

The synthesis method and conditions have an impact on the chemical and physical properties of transition metal nano ferrites. As a result, choosing the right synthesis route is critical for tailoring properties and producing high-quality nano ferrites [30]. Nanosized ferrites may be manufactured using a variety of processes, and the ability to generate nearly any solid solution of nano ferrites opens up the prospect of tailoring their characteristics for a wide range of applications [31]. There are several approaches to identifying and classifying ferrite synthesis methods:

(i) Physical, chemical, and biological, based on the type of processes that occur in the synthesis methods

(ii) Dry and wet methods, based on the presence or absence of a solution

(iii) Conventional and non-conventional, based on their novelty.

Despite numerous categorization approaches, it is difficult to unambiguously categorize the synthesis methods since diverse processes are sometimes combined to produce nanoparticles with specified properties. Many attempts have been made to tailor the size, shape, particle size distribution, surface area, composition, structure, and properties of ferrite nanoparticles by using various synthesis methods or altering synthesis parameters such as annealing temperature and duration, reactant concentration, pH value, stirring speed, doping additives, etc. [32, 33]. Ferrites have been synthesized using a variety of physical and chemical processes [34, 35]. Some synthesis processes are high-energy, complicated operations that need a high processing temperature and considerable reaction time to complete crystallization, as well as the usage of reduction agents that have significant environmental consequences [36, 37]. The relevance of these parameters varies with approach, making repeatability of desired qualities challenging to attain [38]. Another challenge that arises in the majority of synthesis techniques is nanoparticle aggregation after production, which restricts size, shape, and function control [39]. The advantages of wet-chemical synthesis are numerous. On the other hand, single-phase ferrite, can only be formed following high-temperature annealing and is followed by particle growth, aggregation, and coarseness of nanoparticles [40, 41].

2.1 Dry synthesis methods

2.1.1 Combustion method

Because there are no intermediary breakdown or calcination phases, the combustion approach is simple, quick, and economical. This approach makes use of an exothermic, self-sustaining chemical reaction between metal ions and an organic reducing agent (glycine, urea, citric acid, sucrose, hydrazine, and polyvinyl alcohol) [40, 42]. The energy or flame temperature produced during combustion governs the powder properties (i.e., crystallite size, surface area), which is dependent on the kind of fuel and fuel-to-oxidant ratio. Aside from playing a critical influence in the shape of nanoparticles, fuel also impacts phase formation [43]. The heat required to continue the chemical reaction is produced by the process itself, rather than by an external source [44]. Because of their great solubility in water and ease of burning after combining with a suitable fuel, nitrates are the preferred metal precursors. Ammonium nitrate is utilized as an additional oxidant in the combustion reaction, causing the microstructures to expand and, finally, increasing the surface area of the nanoparticle without affecting the proportion of the other participants [45, 46]. Because of their great solubility in water and ease of burning after combining with a suitable fuel, nitrates are the preferred metal precursors. Ammonium nitrate is utilized as an additional oxidant in the combustion reaction, causing the microstructures to expand and, finally, increasing the surface area of the nanoparticles without affecting the proportion of the other participants [45].

This method is a variant of the citrate process, in which ammonium salts and citric acid solutions at pH 7 are dried on a heated plate, resulting in a self-propagating breakdown. As the citric acid polymerized and released CO_2, the cations were transformed to α-Fe_2O_3 and $BaCO_3$, resulting in a strongly exothermic reaction that spread across the whole sample in 20 seconds. The combustion of NH_4NO_3 generated during the neutralization of the solution was the driving force behind this high exothermic reaction, which reached a temperature of 227°C [47]. The rapidity of the reaction assured uniformity, and the development of gas resulted in a porous foam structure that was powdered into a broad agglomeration. The sample was largely $BaFe_{12}O_{19}$ with some α-Fe_2O_3 present after calcining at 700°C and was produced 100% $BaFe_{12}O_{19}$ at 1000°C, although the material exhibits weak magnetic characteristics, likely due to the tiny grain size resulting in poor magnetic ordering [48]. By combining up to 30% wt. of 5 nm carbon NPs (80 m^2g^{-1}) with iron oxide and $BaCO_3$, both of which have particle sizes of less than 50 nm, carbon combustion synthesis of oxides (CCSO) was recently employed to produce $BaFe_{12}O_{19}$ [49]. The reactants were supplied a flow of pure O_2 at 10l min^{-1} after drying and mixing, and the combustion was started using an electrically heated coil. The propagation was unstable with 5% wt. carbon and the front extinguished after 5 mm, but with 6.5-30% wt. carbon, the propagation was steady throughout the reactants and reached maximum temperatures of 900–1200°C. Huang et al., [50] fabricated barium hexagonal ferrite nanoparticles by combustion technique and calcined at two different temperature 700°C and 1200°C as shown in Fig. 1(a). Pure $BaZn_2Fe_{16}O_{27}$ phases are detected after calcining at 1200°C for 2 h as shown in Fig. 1(b). TheXRD patterns fit well with all peaks of the hexagonal-magnetoplumbite structure, as described in JCPDS card number: 52–1868. The tetrahedral M-O link in $ZnFe_2O_4$ and Fe_3O_4 causes considerable absorption at 565 cm^{-1} after heating at 700°C. The C–O asymmetric stretching vibration in $BaCO_3$ is responsible for the absorption at 1430 cm^{-1}.

All of the distinctive peaks from the organic chemicals vanish when the material is calcined at 1200°C, leaving only two primary bands at 570 and 430 cm^{-1}. The spontaneous vibrations of tetrahedral and octahedral M-O bonds in the crystal lattices of the $BaZn_2Fe_{16}O_{27}$ ferrite are responsible for these two different bands(Fig. 1(c)) [50].

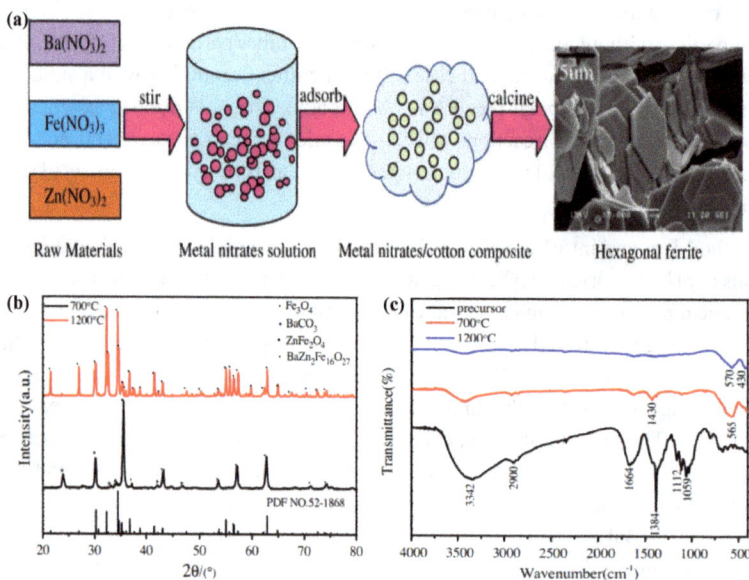

Figure 1:(a) Synthesis of $BaZn_2Fe_{16}O_{27}$ by combustion method,(b) XRD of $BaZn_2Fe_{16}O_{27}$ at two different temperature,(c) FT–IR spectra of $BaZn_2Fe_{16}O_{27}$ calcined at 700°C and 1200°C [50] (Copyright permission with the license number: 5218600522543)

This process takes less time and generates pure and homogenous NPs with no waste, although it does so at a high temperature [51]. The elimination of undesired contaminants as volatile species at high temperatures is credited with the excellent purity of materials generated by the combustion process [36].

2.1.2 Solid-state method

The solid-state approach is the most widely utilized method for producing ferrite powders. The materials are commonly generated in this method by dry or wet milling the initial substances (mainly oxides, oxyhydroxides, or hydroxides), followed by Calcination at a high temperature (800°C or more). The heat treatment promotes the dispersion of constituent raw material, and the solid-state reaction produces crystallite development. The calcined goods are occasionally crushed and ground again to achieve improved uniformity. The powders are sintered after that to achieve the appropriate forms. A protracted heating period at a high temperature, on the other hand, prevents the creation of nano-ferrite powders and restricts the chemical composition to be achieved.

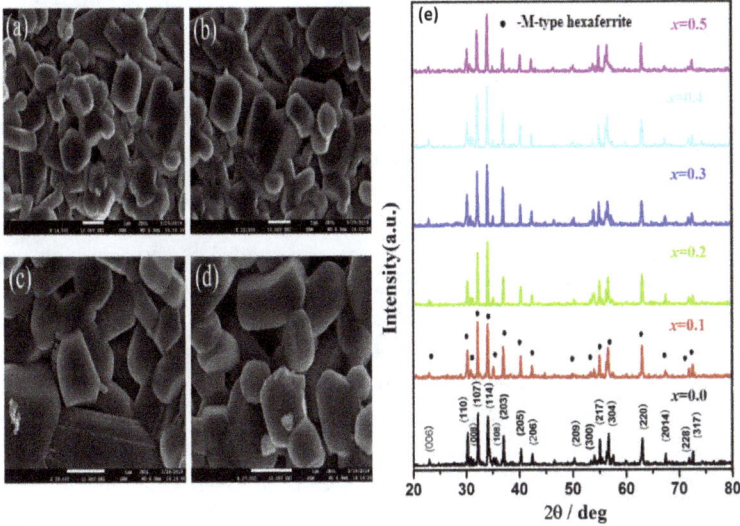

Figure 2: (a-d) SEM and (e) XRD image of $Ba_{0.5}Ca_{0.5}Fe_{12-2x}Mg_xTi_xO_{19}$ hexaferrites [52] (Copyright permission with the license number: 5218620011073)

The contact area between the aggregated particles is restricted for similar magnetic composite particles with various morphologies, resulting in weaker exchange-coupled activity and weaker magnetic characteristics. The solid-state process is commonly used to produce M-type hexagonal ferrites as a typical chemical way to synthesize powders [53, 54]. For the manufacture of hexagonal ferrite nanoparticles, Kitahata et al. employed α-Fe_2O_3, carbonate salts of Ba, Sr, Pb, or Ca, and Fe as raw materials. The materials were sintered for 24 hours at temperatures ranging from 800°C to 1000°C after the combination was ground. The prepared nanomaterials are needle-shaped, with a diametric range of fewer than 0.5 microns. The produced powder has a reasonable coercive force (200-2000 Oe), making it ideal for high-density recording devices.

A solid state reaction technique was used to effectively synthesis single phased M-type hexaferrites $Ba_{0.5}Ca_{0.5}Fe_{12-2x}Mg_xTi_xO_{19}$ ($0.0 \leq x \leq 0.5$). According to the XRD results, all of the synthesised samples are single magnetoplumbite phases with no impurity phases(Fig. 2(e)).The particles have a hexagonal shape, according to their morphology (platelets - like). Platelets-like grains have developed into hexagonal shapes, as can be seen Fig. 2(a-d) [52].

2.2   Wet synthesis method

2.2.1 Co-precipitation method

The co-precipitation method is one of the most widely used methods due to its high yield and ease in generating high purity ultrafine magnetic nanostructured ferrites, the co-precipitation process is one of the most extensively utilized procedures [55]. The co-precipitation technique is a simple and cost-effective approach that allows for easy particle size and composition control, uses little heat, and produces materials with excellent crystallinity, homogeneity, and good textural qualities [55]. Co-precipitation technologies have several limitations, including severe agglomeration, poor crystallinity and particle size dispersion, and the need for pH regulation [56, 57]. In this situation, inorganic salts (chloride, sulfate, and nitrate) are dissolved in water or solvents to produce homogenous solutions. The precipitate is collected by filtration or centrifugation, washed, and dried after pH adjustment in the range of 7–12 under continuous stirring. Particle aggregation and crystal growth are caused by the rate of pH change [55]. Using Co^{2+} and Fe^{3+} salts in the presence of a strong base is the most frequent strategy to produce nanostructured ferrites via chemical co-precipitation [39]. The size, shape, and magnetic characteristics of nanostructures may be modified by modifying experimental parameters such as reaction temperature and duration, reagents feeding rate and concentration, pH, drying temperature, and so on [57]. Ferrites are made from aqueous solutions of chlorides, nitrates, or iron (III) sulphates, as well as divalent metals like Ni, Co, Ba, Mg, Sr, and others, at concentrations sufficient for ferrite composition and concomitant precipitation as hydroxides in an alkaline media (NaOH) [58, 59]. Since the early 1960s, ferrites have been prepared via chemical coprecipitation of salts with hydroxide, resulting in a precipitate having all of the components combined at an ionic level [60, 61]. Haneda et al. [62] showed that an iron-deficient non-stoichiometric mixture must be employed instead of the right ratio of 12, for example with a Fe:Ba ratio of 10–11 for $BaFe_{12}O_{19}$. To produce a pure M-phase, the Fe:Ba ratio of the $SrFe_{12}O_{19}$ must be smaller than 9 [63]. $BaFe_{12}O_{19}$ with a submicron particle size may be generated using this technique between 750 and 900°C, and its density can be enhanced if washed by centrifugation rather than decantation [63]. To create the single-phase product in coprecipitated $BaFe_{12}O_{19}$, an iron-deficient non-stoichiometric mixture is

Materials Research Foundations **142** (2023) 35-65 https://doi.org/10.21741/9781644902318-2

generally required, with a Fe:Ba ratio of 10 to 11, while ratios as low as 8 have been described as optimal [64, 65]. As water is lost and the hydroxides are converted to oxides, the FeOOH and $Ba(OH)_2$ coprecipitates and dried display an endothermic between 130°C and 350°C, but the material remains amorphous until $BaFe_{12}O_{19}$ starts to form at 700°C, with no crystalline precursors. As the M-phase crystallizes, Ms begins to rise, and there is a strong exothermic at 760°C correlating to this crystallization. Differential thermal analysis (DTA) uses a heating rate of 10°C/min, which results in unrealistically high-temperature measurements for fast processes like crystallization. At 900 degrees Celsius, all of the material is transformed to $BaFe_{12}O_{19}$, and grain growth occurs at 1000 degrees Celsius. Further, than $BaFe_{12}O_{19}$, no other phases have been reported [66]. Using nitrates and chlorides with NaOH, Lisjak and Drofenik [67] examined the generation mechanism of $BaFe_{12}O_{19}$ from the stoichiometric co-precipitated precursor, which was subsequently calcined between 300°C and 800°C for 5 hours. They discovered that α-Fe_2O_3 crystallizes around 290 degrees Celsius, and that during the process, $BaCO_3$ and a little amount of $Ba(OH)_2$ (but not all) are formed in an air environment. There was no indication of $BaFe_{12}O_{19}$ production below 500°C, but it begins to form above 500°C from amorphous precursors. $BaCO_3$ combines with α-Fe_2O_3 to generate $BaFe_2O_4$, which then reacts with additional α-Fe_2O_3 to form $BaFe_{12}O_{19}$ at 600 °C and higher. This reaction was so quick that no $BaFe_2O_4$ was left after 30 minutes at 700°C. Lisjak and Drofenik [68] also looked into using ethanol as a solvent instead of water during coprecipitation to prevent concerns with absorbed CO_2 from the atmosphere producing $BaCO_3$ in the water. In both air and Ar flow, they employed stoichiometric iron ratios for chloride and acetate salts, which were precipitated with NaOH in ethanol: water (3:1) solution. It was discovered that when chloride precursors were coprecipitated, $BaFe_{12}O_{19}$ crystallized below 500°C/10 h, despite the presence of crystalline α- and γ-Fe_2O_3, and that the pure $BaFe_{12}O_{19}$ phase formed at 600 and 700°C/10 h for the samples coprecipitated in Ar and air, respectively.Hu et al., [69] prepared barium ferrite by coperecipitation method. The XRD data showed that the generation of barium hexaferrite and the α-Fe_2O_3 phase happens at the same time over a reasonably short co-precipitation reaction period (3 h or 4 h), followed by calcining at various temperatures (Fig. 3(d-f)). Meanwhile, increasing the calcining temperature (700-1100°C) promoted $BaFe_{12}O_{19}$ crystallisation and resulted in the single phase.

The reaction duration of co-precipitation and the calcination temperature are both important factors in the effectiveness of $BaFe_{12}O_{19}$ particles. The particle size improved as the calcination temperature is raised, while the co-precipitation reaction time remained constant. Meanwhile,the calcination temperature remained constant, the particle size grew as the co-precipitation reaction time grew.

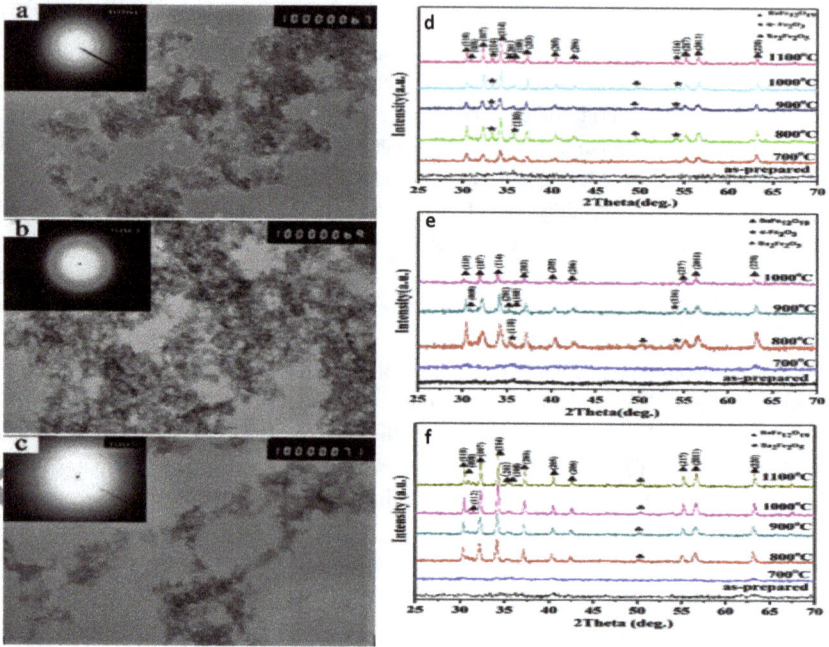

Figure 3: (a-c) TEM and (d-f) XRD images for precursors co-precipitated for 3 h (a,d), 4 h (b,e), 5 h (c,f) [69] (Copyright permission with the license number: 5218621454407)

2.2.2 Sol-gel method

The sol-gel technique is a low-temperature process that involves metal precursors (salts or alkoxides) being hydrolyzed and condensation reactions taking place, resulting in the development of a three-dimensional inorganic network [70]. Sol is made by converting monomers into a colloidal solution, whereas gel is made by connecting particles into a network after the solvent has evaporated. The sol-gel technique for preparing nanocomposites (NCs) is a simple, low-cost, and environmentally friendly technology because it provides for precise control of the microstructure, particle size, dispersion, structure, and chemical content by carefully monitoring the production conditions [71, 72]. The sol-gel approach has been used to make ferrite NPs that are very fine, dense, homogeneous, and single-phase. In comparison to other traditional procedures, the sol-gel exhibits strong stoichiometric control and permits ferrites to be produced at low temperatures. The nanomaterials produced can be made into films or colloidal powders. The primary drawback is that the synthesis process is inefficient and takes a long time [73].

The sol-gel procedure was used to create fine-grained polycrystalline ferrites with a limited size distribution. The process involves mixing precursor particles (inorganic or Metallo-organic on the colloidal scale) that are maintained when the mixture is condensed to a gel. In aqueous sol-gel synthesis, a base co-precipitates aqueous metal salt solutions, which are then washed, dried, and treated to create a colloidal sol, which may then be concentrated to a gel and frozen to generate ferrite [74]. To make the $BaFe_{12}O_{19}$, Mccolm and Clark [75] used $Fe(OC_3H_7)_3$ and $Ba(OC_3H_7)_2$, as well as C_3H_7OH as a solvent. The alcogel was aged for 12-24 hours after adding water. To make an amorphous powder, the particles were separated by centrifugation and dried at 100°C. Calcination at 700°C for 2 hours yielded $BaFe_{12}O_{19}$ with a particle size of around 1m. According to Surig et al. [76], if a sol includes precipitated $Ba(OH)_2$, which is particularly stable at a high pH, the sol will be very basic, and hence a sol cannot be generated by acid digestion and peptization. To make a sol, an organic coordinating agent like ethylene glycol is commonly added to the hydroxide solution, which forms a gel structure after the water evaporates. A non-stoichiometric combination was developed when $BaFe_{12}O_{19}$ was obtained using a sol-gel process utilizing the evaporation of glycol containing co-precipitated salts to give a homogeneous gel, and a Fe:Ba ratio of 10:5 produced the M- ferrite at 900 °C/1h with a grain size of 200 nm. Only the barium-rich precursor will generate pure M-ferrite without any α-Fe_2O_3 as a secondary phase, but with a $BaFe_2O_4$ ratio of less than nine as a secondary phase, showing that the production of $BaFe_{12}O_{19}$ by this approach has a tight composition window. The presence of the nitrate anion was shown to be a key factor in the synthesis of the intermediate phases α-Fe_2O_3 and $BaFe_2O_4$ throughout a wide range of temperatures when the Fe/Ba ratio was set at 11.6, and $BaFe_{12}O_{19}$was generated from nitrate or hydroxide sol-gel precursors. The non-magnetic phases were eliminated by heating the gel at 450°C for 5 hours to remove any organic compounds, with $BaFe_{12}O_{19}$ being formed at 750°C, however, the best magnetic characteristics were obtained at 950°C [47]. Pullar et al. [77, 78] demonstrated that organic-free $BaFe_{12}O_{19}$ and $SrFe_{12}O_{19}$ may be synthesized from stoichiometric precursors in an aqueous sol–gel-based technique. Except in the instance of $BaFe_{12}O_{19}$ generated from a stabilized halide sol, when crystallized barium halide salts were also detected, no $BaFe_2O_4$, γ-Fe_2O_3, or any other ferrite intermediary was observed in the M-ferrite precursor powders, with M-ferrites forming straight from α-Fe_2O_3 [79]. This revealed that the creation of γ-Fe_2O_3 or $BaFe_2O_4$ is not required for the formation of M-ferrites, however, there was some speculation that the presence of halides could have slowed the development of these precursors. Although halide based sol precursors were more stable, it was also seen that the halides persisted until high temperatures (~10 % wt at 600 °C), particularly chloride ions, delaying the onset of $BaFe_{12}O_{19}$ formation until 750/800 °C for $SrFe_{12}O_{19}$/$BaFe_{12}O_{19}$, and the single-phase M-ferrites were not obtained until 900/1000 °C, because in this point there was no halide remaining in the powder.

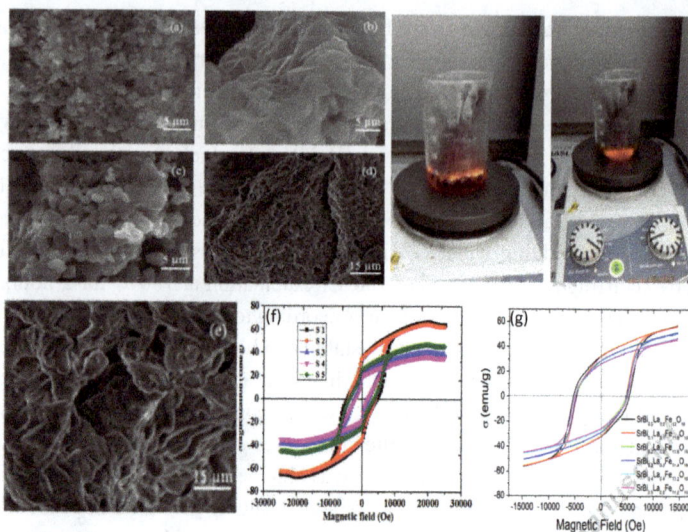

Figure4: (a-e) shows the SEM images of Ce-Zn doped M-type SrFe$_{12}$O$_{19}$ hexagonal ferrites, (f) M-H curves of Ce-Zn doped M-type SrFe$_{12}$O$_{19}$ hexagonal ferrites[80](Copyright permission with the license number: 5221180418063)and (g) M-H curves of SrBi$_x$La$_x$Fe$_{12-2x}$O$_{19}$ (0.0 ≤ x ≤ 0.5) hexaferrites [81] (Copyright permission with the license number:5221180282480)

As the halides were driven off, this resulted in a weight loss of 4–5% between 800 and 1000 °C. In halide-free nitrate sol precursors, the ferrites formed immediately from α-Fe$_2$O$_3$ at lower temperatures, such as 650 °C for BaFe$_{12}$O$_{19}$, and became the sole phase at 750 °C, with no crystalline nitrate precursors seen. The transition from α-Fe$_2$O$_3$ at 650 °C to single-phase SrFe$_{12}$O$_{19}$ at 700 °C was even quicker in halide-free SrFe$_{12}$O$_{19}$ [79], demonstrating that γ- Fe$_2$O$_3$ or BaFe$_2$O$_4$ are not required for the creation of the M phase. Sol-gel auto-combustion is used to make M-type Sr-hexaferrites with the chemical composition Sr$_{1-x}$Ce$_x$Fe$_{12-y}$Zn$_y$O$_{19}$ (x = 0.000, 0.025, 0.050, 0.075, 0.10, y = 0.00,0.025, 0.50, 0.75, 1.0). Obtained materials are single phase, according to X-ray diffraction. SEM investigation demonstrates hexagonal platelet grains with grain sizes ranging from 1.329 to 0.856 m. There is also some agglomeration, which is caused by magnetic behaviour and weak Vander Walls forces (Fig. 4 (a-e)) and (Fig. 4 (f)) [80]. On the other side Auwal et al., fabricated SrBi$_x$La$_x$Fe$_{12-2x}$O$_{19}$ (0.0 ≤ x ≤ 0.5) ferrites were prepared by sol-gel method. The ferromagnetic nature of SrBi$_x$La$_x$Fe$_{12-2x}$O$_{19}$ ferrites at RT is demonstrated by the unique characteristics of hysteresis curves (Fig. 4 (g)). As the non-magnetic Bi and La concentrations increase, the σs of NPs decreases from 63.45 emu/g to 49.85 emu/g [81].

2.2.3 Spray pyrolysis method

Spray pyrolysis entails turning the reagent mixture into aerosol droplets, solvent evaporation, solute condensation, and drying, followed by high-temperature thermolysis of the particles [38]. By separating the fluid into droplets, this approach enables control of the particle formation environment [82]. It is often ideal for the manufacture of mixed metal ferrites because it achieves perfect stoichiometry retention on the droplet size and a relatively homogenous component distribution [82]. This approach permits the synthesis of a wide range of hollow or porous particles with possible uses in thermal insulation or catalyst support by regulating the kind of thermolysis reaction, the type of precursors, the gaseous carrier, deposition duration, and substrate temperature [83].

This procedure entails the evaporation of the solvent at high pressure to precipitate a concentrated solution of cations that will form the ferrite. The solution is sprayed into minute droplets at high pressure, and the solvent is swiftly evaporated by an upward flow of hot gas. Tang et al. [84] generated fine hollow spheres of $BaFe_{12}O_{19}$ by aerosolizing a metal nitrate solution that was atomized and sprayed in a N_2 gas flow into a heated zone for about one second, following which the mostly amorphous spheres were recovered.

2.2.4 Microwave-assisted combustion method

To synthesis NPs, the microwave-assisted combustion approach uses microwave radiations to bombard the reaction solution. Unlike traditional heating technologies, where heat is delivered from the outside, microwave radiation is absorbed and transformed to thermal energy, resulting in heat being created inside a material. This heating provides for significant time and energy savings during processing [85]. Simple, homogenous nucleation, short reaction time, high production rates, environmental friendliness, outstanding repeatability, cheap energy cost, easy handling, and parameter control are the key benefits of this approach over other synthesis methods. Due to the interaction of microwaves, the reagents are combined at the molecular level in the microwave-assisted combustion process, allowing for good control of stoichiometry, purity, homogeneity, and morphology [86]. However, the approach is costly, owing to the expensive fuels necessary to stimulate and manage the combustion process in line with propellant chemistry principles, such as urea, glycine, L-alanine, carbohydrazide, or citric acid, and is unsuitable for scale-up and reaction monitoring [86, 87]. The necessary heat is provided by microwaves in the microwave hydrothermal process, which has the advantage of very quick heating to a particular depth, allowing the production of homogeneous nanomaterials with small particle size distribution. Nitrates of were dissolved in deionized water at a pH of 9.4. The mixture was sealed with tetrafluorometoxil (TFM) and microwaved for 30 minutes at 160 degrees Celsius.

Figure 5: (a-f) SEM micrograph of strontium hexaferrite nanoparticles [88] (Copyright permission with the license number: 5221200957450)

The resultant wet mixture was dried before being sintered for 30 minutes using polyvinyl alcohol (PVA) as a binder. This approach allows for quicker healing, lower costs, and the production of very fine, uniform NPs [86]. A microwave-assisted technique was used to make strontium hexaferrite nanoparticles. SEM was used to monitor the changes in particle shape caused by annealing (Fig. 5). The platelet shape did not change much at annealing temperatures up to 800°C. The platelet shape was intact after annealing at 1000°C, but the platelet thickness increased (Fig. 5(d)) due to the start of sintering. Above 1050°C, the particles appeared completely sintered (Fig. 5(e-f)), with significantly increased Dm and Tm [88].

An Introduction to Hard Ferrites: From Fundamentals to Practical Applications   Materials Research Forum LLC
Materials Research Foundations **142** (2023) 35-65   https://doi.org/10.21741/9781644902318-2

2.2.5 Microemulsion method

The microemulsion technique involves combining reactant-containing microemulsions, forming microdroplets, and trapping fine aqueous microdroplets inside surfactant molecule assemblies. During the nanoparticle creation process, this approach causes a locking up effect, which inhibits the growth dynamics, particle nucleation, and agglomeration. The main benefit is the ability to obtain monodispersed NPs with a variety of morphologies, whereas the main negatives include low efficiency and difficulties scaling up [89]. A microemulsion is a dispersion of two immiscible liquids stabilized by a surfactant interfacial coating, resulting in 5–10 nm domains of one liquid in the other. Aqueous droplets are scattered in an organic solvent and constantly collide, agglomerate, and break apart, mixing and trading their solute contents. When two micro-emulsions are combined that are similar except for the fact that they contain distinct reactants in each aqueous phase, a reaction can occur, resulting in nanoparticles, and the shape can typically be controlled better than in regular coprecipitation [90]. Rawlinson and Sermon [91] used this method of controlled precipitation to synthesize $BaFe_{12}O_{19}$ by using a stoichiometric metal salt solution and an ammonium carbonate/sodium hydroxide solution as the two aqueous phases, resulting in a nanosized iron–barium–carbonate/hydroxide precipitate that was then separated, washed, and dried. $BaFe_{12}O_{19}$ forms at 600°C and becomes a single phase at 925°C, with grain size control in the 120–170 nm range.

2.2.6 Citrate precursor method

The citrate precursor technique is a wet chemical process that includes combining aqueous solutions of precursor salts (metal nitrates) with an aqueous solution of citric acid, then heating at 80 degrees Celsius and annealing at 700 degrees Celsius [92]. The precursors are thermally degraded into ferrite powders during this procedure [93]. The advantages of this process over other chemical methods include its high reactivity, short reaction time, low synthesis temperature, homogeneous ion distributions, and inexpensive cost [94]. Ultrafine particles may be made from degraded citrates at low temperatures because the breakdown reaction is extremely exothermic, resulting in the production of CO_2 and a porous product with a large surface area. The Pecchini procedure is another name for this approach. Sankaranarayanan et al. [95] synthesized $BaFe_{12}O_{19}$ from a stoichiometric solution of metal salts and citric acid in a cation: citrate = 1 ratio, with ammonia added to elevate the pH and make a homogeneous solution. This was heated to 80 °C to remove any leftover ammonia, then ethanol was added to produce an iron/barium/citrate complex, which precipitated owing to alcohol dehydration, and then dried and decomposed at 425–470 °C for 48 hours. This amorphous product, which had a grain size of less than 10 nm, was heated to 600 °C to produce $BaFe_{12}O_{19}$, albeit a temperature of over 700 °C was needed to generate a completely crystalline sample with acceptable magnetic characteristics and a grain size of 60–80 nm. At 800 °C, nanocrystalline $SrFe_{12}O_{19}$ particles with a diameter of 42 nm have also been produced [96]. Sankaranarayanan et al. [95] used a stoichiometric amorphous precursor with 10 nm, decomposed at 420-470 °C, and produced a single crystalline phase of hexaferrite after 550 °C, with the grain size of 60 nm in the XRD pattern at 700° C. Mössbauer investigations showed that non-magnetic $BaFe_2O_4$ may exist and that at this temperature, certain cations had not diffused entirely into the lattice, but about 800° C, Mössbauer results agreed with XRD that the pure M phase had formed, and the grain size was below 10 nm [96]. For citrate routes, there are several contradicting reports on the optimal Fe:Ba and citrate: metals ratios. The optimum ratios for $BaFe_{12}O_{19}$ formation from a citrate route, according to Li et al. [97], were a stoichiometric Fe:Ba of 12, and citric acid: iron(III) nitrate of 3 (solution pH = 9), which decompose at 208 °C, begin to form $BaFe_{12}O_{19}$ at 750 °C, and produce single-phase $BaFe_{12}O_{19}$ with good magnetic properties at 800 °C. In a study of the influence of pH and citric acid: metals ratio on the synthesis of $BaFe_{12}O_{19}$, it was discovered that $\gamma-Fe_2O_3$ and $BaCO_3$ were always recognized as combustion products, with $\alpha-Fe_2O_3$ appearing on occasion, but never $BaFe_{12}O_{19}$. Therefore the pH >3, and with barium as well as pH >7, full complexationbetween citrate and iron was obtained, and citrate: metals ratio of 1.5 was optimum. With a citrate ratio of 1.5, pH = 9, and calcination at 800°C/4 h, the best product was produced [98].

According to Huang et al. [99], at a pH of 7, the optimal Fe:Ba ratio is 11.5, and the citrate: metals ratio is 2:1.The citrate precursor technique is used to make strontium hexaferrite nanoparticles. The iron oxide impurity in the produced hexagonal strontium ferrite at 850 °C is removed by increasing the annealing temperature to 950 °C (Fig. (c)).Fig. 6(a-b) shows high-resolution micrographs of $SrFe_{12}O_{19}$ annealed at two different tempreature i.e. 850 °C and 950 °C. This demonstrates the formation of uneven and agglomerated particles, however it can also be observed that at a higher annealing temperature of 950 °C, particle size is equivalent to structural data, but grain size is better [100].

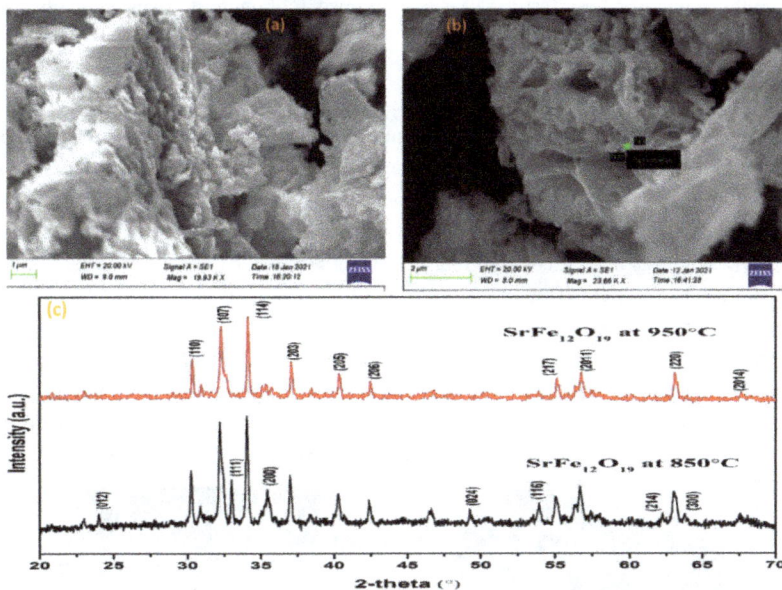

Figure 6: (a-b) High-resolution micrographs of SrFe$_{12}$O$_{19}$ annealed at two different tempreature i.e. 850 °C and 950 °C and (c) XRD pattern of SrFe$_{12}$O$_{19}$ annealed at two different tempreature i.e. 850 °C and 950 °C [99-100] (Copyright permission with the license number: 5221230056201)

2.2.7 Thermal decomposition method

The thermal decomposition technique works by heating reactants (metallic precursors) at various temperatures. Surfactants (oleylamine, oleic acid) are introduced to the decomposition process to regulate the nucleation and development of the NPs and, as a result, enhance the physical characteristics of the material, such as crystallite size, porosity, and specific surface. Surfactants act as a protective envelope around the NPs, preventing coalescence and improving the material's physical qualities including crystallite size, porosity, and specific surface area. The magnetic characteristics may be controlled using these factors [101]. The approach is straightforward, low-cost, uses low reaction temperatures, is environmentally benign, creates highly monodispersed particles with a limited size distribution, and produces no byproducts. The primary disadvantage is the simultaneous removal of surfactants and particle size control [38, 102]. The use of heat

53

treatment to manufacture ferrite NPs is restricted, although it takes less time than alternative synthesis methods.

2.2.8 Hydrothermal method

A solution of metal salts and a base is autoclaved under pressure to produce the product in hydrothermal synthesis. This is usually a mixed phase with unreacted precursors and occasionally α-Fe_2O_3, which may be eliminated by washing with weak HCl. Ataie et al. [103] investigated the effects of NaOH, KOH, $(C_2H_5)_4$NOH, and NH_4OH as bases on $BaFe_{12}O_{19}$ synthesis and discovered that at 220 °C/5 h, NaOH and KOH create micron-sized $BaFe_{12}O_{19}$ platelets. This process was used by Jean et al. [104] to make $SrFe_{12}O_{19}$ with a Fe:Sr ratio of 8. The hydrothermal synthesis of hexaferrite has recently gained popularity as a cost-effective and straightforward method of producing ferrite nanoparticles (NPs). Although the generated product contains hexaferrite NPs, the synthesis temperatures are often lower than 250 °C, resulting in a mixed-phase or chemically heterogeneous product, necessitating annealing to obtain pure phase hexaferrite.

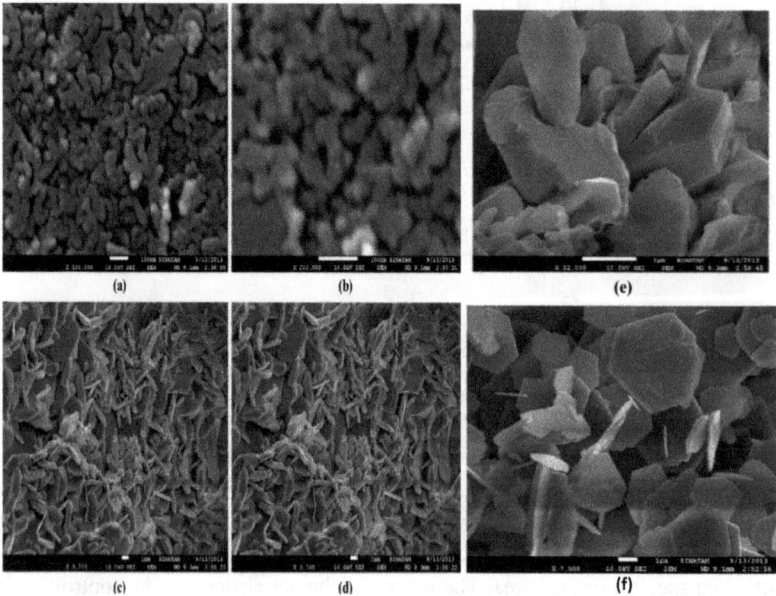

Figure 7: SEM images of the products synthesized at (a) 170°C, (b) 180°C, (c) 190°C, (d) 200°C, (e) 210°C, (f) 220°C [105] (Copyright permission with the license number:5221251098381)

By using hydrothermal synthesis, Lee et al. [106] were able to synthesize a pure phase of $BaFe_{12}O_{19}$ from stoichiometric combinations of iron and barium. However, because the result is weakly crystalline, it must be annealed at 900°C for the Ba^{2+} ions to diffuse entirely into the M lattice. The barium excess causes carbonate impurities in hydrothermal hexaferrite products at ratios below Fe:Ba = 10, since the barium and strontium react with CO_2 in the air during the process. When autoclaved with $Ba(OH)_2$ at 250–280 °C, Liu et al. [107] indicated that a Fe:Ba ratio of 8, an OH^-:NO_3^- the ratio of 2, and a reaction at 230 °C/48 h were required. They also looked at how the OH^-:NO_3^- ratio (1–5), Fe:Ba ratio (8–12), and temperature (200–240 °C) affected the hydrothermal synthesis of $BaFe_{12}O_{19}$ from nitrates with NaOH in this work.

At 230 °C, single-phase $BaFe_{12}O_{19}$ was produced with OH^-:NO_3^- ratios of 2–5, and the $BaFe_{12}O_{19}$ particle size dropped from 1.8 to 1.2 m. These smaller particles produced agglomerates (5–10 m in size) when a ratio of 5 was used, however when a ratio of 2 was used, they formed nicely scattered hexagonal platelets. Pure $BaFe_{12}O_{19}$ formed in a Fe: Ba ratio of 8 (230 °C/48 h), some α-Fe_2O_3 was present at 10, and this impurity became the major phase at 12. As the reaction temperature climbed from 200 to 240 °C/48 h, particle size grew from 1 to 2 m, and as the reaction time was prolonged at 230 °C, $BaFe_{12}O_{19}$ became the main product after 15 h, and pure ferrite was achieved after 25 h– higher temperatures needed less time to achieve this. A hydrothermal technique was used to successfully manufacture $SrFe_{12}O_{19}$ particles. Fig. 4 shows SEM micrographs of all products at various magnifications for various temperatures. The products made at 170, 180, 190, and 210 degrees Celsius lack a plate-like hexagonal shape (Fig. 7). The Fe/Ba mole ratio was set to 8:1, the ideal temperature was established, and it was discovered that above 200 degrees Celsius, an uniform platelet-like structure with a limited size dispersion was achieved [105].

2.2.9 Reverse micelle method

The reverse micelle approach is based on the production of water-in-oil emulsions in which the water-to-surfactant ratio determines the size of water pools in which aqueous chemical syntheses occur. The process entails combining the metal salt precursors and the precipitating agent, then forming microemulsion droplets that operate as a nano-reactor, where nanometric size precipitates are generated as a result of reactant coalescence and droplet collision [109]. This technique is suited for room temperature processes such as the precipitation of oxide NPs because it provides for great control of particle size, size distribution, chemical stoichiometry, and cation occupancy [108]. Metallic chlorides and NaOH were dissolved in two separate microemulsion systems made from water, sodium dodecyl sulfate, 1-butanol, and n-hexane using ultrasonication. The two microemulsions were mixed until metallic hydroxides precipitated, then filtered, washed, dried, and annealed at 400 degrees Celsius. The particle size of ferrites is greatly influenced by the composition of the surfactant, the water to surfactant ratio, and the pH value of the solution, all of which have an impact on their characteristics [109].

2.2.10 Polyol method

Under reflux circumstances, the polyol approach involves reducing metallic oxide or metallic complexes using a high boiling point solvent that serves as both a solvent and a reducing agent for the metallic ions. The method uses a suspension of precursors in liquid polyol (ethylene glycol, diethylene glycol, and triethylene glycol), followed by heat treatment up to the boiling point of the polyol, to regulate particle development and prevent agglomeration. The key benefit of this technology is that it produces uniformly sized soft and hard magnetic NPs, whereas the main drawbacks are the high temperature and extended processing time [110, 111]. Many oxides and NCs materials have been synthesized using the polyol approach.

2.2.11 Spray drying method

The ferrite slurry droplets are sprayed into a hot air vertical evaporation tube in this manner. The primary disadvantages are the inflating faults, which appear as big spaces inside the granule, and the formation of capillary tension as a result of the fast evaporation of water. This stress may cause unwanted diffusion and segregation, reducing the compositional and morphological homogeneity of NPs [112, 113]. Because flaws may resist annealing, the spray droplets can be frozen with liquid nitrogen and freeze-dried to maintain the desired particle size and prevent defects. There are no unwanted capillary tensions during water sublimation, which result in hard, unbreakable aggregates and voids, resulting in a stiff porous product of loose, non-agglomerated particles [114].

2.2.12 Sonochemical method

One of the most promising methods for obtaining ferrite NPs is the sonochemical approach. The acoustic cavitation phenomenon, which involves the development, growth, and collapse of bubbles in liquid media, gives birth to sonochemistry. The intense reaction conditions (high temperature (5000 K), pressure (20 MPa), and cooling rate (1010 K/s) result in a variety of distinct features in the generated particles [115]. In a one-step, surfactant-free sonochemical approach, highly crystalline, monodisperse $CoFe_2O_4$ NPs with uniform spherical shape and high MS values were synthesized. Furthermore, the synthesis time was short (about 70 minutes), and no additional annealing was required. Simple, low-cost, safe, environmentally friendly, uniform size distributions, large surface area, short response time, and superior phase purity are some of the advantages of this technologyover traditional approaches [115]. The low concentration of produced NPs and particle agglomeration are disadvantages.

3. Comparision of synthesis methodologies

The distribution of metallic ions among crystallographic lattice sites, which is susceptible to the synthesis process used, has a significant effects on the properties of ferrites materials. The approach to choose is mostly determined by the applications and qualities that are needed. Ferrite nanoparticles are generated using several chemical processes at various temperatures and sintering conditions, allowing for a comparison of the structural and magnetic properties of ferrites produced using various synthesis techniques. As we know, the classical ceramic method for producing barium ferrites involves high-temperature solid-state reactions between the constituent oxides and carbonates [62, 116]. The disadvantages of this conventional method also include rather large and non-uniform particle size and introduction of the impurities, which restrict further improvement in the performance of the products. To overcome these difficulties and to meet the requirements for new applications, various wet-chemical processes like sol-gel citrate [117, 118], hydrothermal method [119], micro-emulsion process [119], co-precipitation technique [68], and plasma spraying [120]. have been considered for the production of barium ferrite ultrafine powders with excellent magnetic properties.Reverse micelle approach is suited for room temperature processes such as the precipitation of oxide NPs because it provides for great control of particle size, size distribution, chemical stoichiometry, and cation occupancy [108]. The particle size of ferrites is greatly influenced by the composition of the surfactant, the water to surfactant ratio, and the pH value of the solution, all of which have an impact on their characteristics [109]. In case of thermal decomposition technique surfactants also act as a protective envelope around the NPs, and improving the material's physical qualities including crystallite size, porosity, and specific surface area. The magnetic characteristics may be controlled using these factors [101].

The magnetic characteristics of hexaferrite are influenced by the crystallographic phase composition, which is influenced by the synthesis technique. Cernea et al., were synthesized $Ba_xSr_{1-x}Fe_{12}O_{19}$, where x = 0.05, 0.15, 0.25, 0.35, using two methods: sol-gel and solid-state reaction. SEM data of the ferrites demonstrate that grain size reduces as x increases, however, X-ray diffraction results point to a pure hexaferrite phase in sol-gel experiments and a low level of impurity (5-10 molar percent) in solid-state reaction experiments. Mössbauer spectroscopic findings show that the sol-gel technique samples have a lower occupancy of the Fe specific sites than the solid-state method samples, which has a direct impact on the particular magnetic characteristics.

Conclusion

In this chapter, we've attempted to summarise the fundamentals of ferrites, including classifications. The ferrite synthesis process used has an impact on the size, structure, microstructure, magnetic behaviour, electrical conductivity, and other features of a desired product. The process and circumstances used to manufacture ferrite have a big impact on its properties. As a result, the goal of this study was to show different methods for making nano-ferrites. Furthermore, the microstructure characteristics of the sample, such as grain sizes and specific surface area, are influenced by the preparation procedures.

References

[1] P. Dhiman, N. Dhiman, A. Kumar, G. Sharma, M. Naushad, A.A. Ghfar, J. of Mol. Liq.294, 111574 (2019). https://doi.org/10.1016/j.molliq.2019.111574

[2] J. Went, Phil. Tech. Rev. 13, 361(1952). https://doi.org/10.2307/371513

[3] M. Weiss, P. Anderson, Phys. Rev.98, 925(1955). https://doi.org/10.1103/PhysRev.98.925

[4] T. Sebastian, Y. Ohdaira, T. Kubota, P. Pirro, T. Brächer, K. Vogt, A. Serga, H. Naganuma, M. Oogane, Y. Ando, Appl. Phys. Lett. 100, 112402(2012). https://doi.org/10.1063/1.3693391

[5] K. Sixtus, J. kronenberg and RK Tenzer, J. appl. Phys, 27 (1956) 1051. https://doi.org/10.1063/1.1722540

[6] Y.-Y. Song, C.L. Ordóñez-Romero, M. Wu, Appl. Phys. Lett. 95, 142506 (2009). https://doi.org/10.1063/1.3246170

[7] V.P. Singh, R. Jasrotia, R. Kumar, P. Raizada, S. Thakur, K.M. Batoo, M. Singh, J. condens. matter phys.8, 36 (2018). https://doi.org/10.4236/wjcmp.2018.82004

[8] Y. Bai, J. Zhou, Z. Gui, L. Li, J. magn. magn. mater.278 208-213 (2004). https://doi.org/10.1016/j.jmmm.2003.12.1389

[9] A. Gadkari, T. Shinde, P. Vasambekar, Mater. Chem. Phys.114, 505-510 (2009). https://doi.org/10.1016/j.matchemphys.2008.11.011

[10] P. Dhiman, A. Kumar, M. Shekh, G. Sharma, G. Rana, D.-V.N. Vo, N. AlMasoud, M. Naushad, Z.A. ALOthman, Environ. Res.197, 111074 (2021). https://doi.org/10.1016/j.envres.2021.111074

[11] P. Pramanik, A. Pathak, Bull. Mater. Sci.17, 967-975(1994). https://doi.org/10.1007/BF02757573

[12] S. Kumar, D.P. Dubey, S. Shannigrahi, R. Chatterjee, J. Alloys Compd.774, 52-60(2019). https://doi.org/10.1016/j.jallcom.2018.09.339

[13] S. Arcaro, J. Venturini, (Springer, 2021).

[14] R.C. Pullar, PProg. Mater. Sci.57, 1191-1334 (2012) https://doi.org/10.1016/j.pmatsci.2012.04.001

[15] A. Goldman, Modern Ferrite Technology, 151-216(2006).

[16] M.W. Pieper, A. Morel, F. Kools, J. Magn. Magn. Mater. 242, 1408-1410 (2002). https://doi.org/10.1016/S0304-8853(01)00963-5

[17] G.F. Dionne, (Springer, 2009).

[18] Ü. Özgür, Y. Alivov, H. Morkoç, J. Mater. Sci. Mater. Electron.20, 789-834(2009). https://doi.org/10.1007/s10854-009-9923-2

[19] F. Lotgering, P. Vromans, M. Huyberts, J. Appl. Phys.51, 5913-5918(1980). https://doi.org/10.1063/1.327493

[20] A. Paoluzi, F. Licci, O. Moze, G. Turilli, A. Deriu, G. Albanese, E. Calabrese, J. Appl. Phys.63, 5074-5080(1988). https://doi.org/10.1063/1.340405

[21] D. Samaras, A. Collomb, S. Hadjivasiliou, C. Achilleos, J. Tsoukalas, J. Pannetier, J. Rodriguez, J. Magn. Magn. Mater.79, 193-201(1989). https://doi.org/10.1016/0304-8853(89)90098-X

[22] S. Ruan, B. Xu, H. Suo, F. Wu, S. Xiang, M. Zhao, J. Magn. Magn. Mater.212, 175-177 (2000). https://doi.org/10.1016/S0304-8853(99)00755-6

[23] P.B. Braun, Philips Res. Rep. 12 491-548(1957). https://doi.org/10.2307/1524539

[24] F. Leccabue, R. Panizzieri, G. Bocelli, G. Calestani, C. Rizzoli, N.S. Almodovar, J. Magn. Magn. Mater.68, 365-373 (1987). https://doi.org/10.1016/0304-8853(87)90015-1

[25] P. Dhiman, R. Jasrotia, D. Goyal, G.T. Mola, Ferrite: Nanostructures with Tunable Properties and Diverse Applications 112, 336 (2021). https://doi.org/10.21741/9781644901595-10

[26] H. Elkady, M. Abou-Sekkina, K. Nagorny, Hyperfine Interact.128, 423-432 (2000). https://doi.org/10.1023/A:1012612405813

[27] H. Zhang, J. Zhou, Z. Yue, P. Wu, Z. Gui, L. Li, Mater. Lett.43, 62-65(2000). https://doi.org/10.1016/S0167-577X(99)00231-1

[28] X. Wang, L. Li, S. Su, Z. Gui, Z. Yue, J. Zhou, J. Eur. Ceram. Soc.23, 715-720 (2003). https://doi.org/10.1016/S0955-2219(02)00157-7

[29] D.L. Kunwar, D. Neupane, J.N. Dahal, S.R. Mishra, Adv. Mater. Sci. Eng.9, 175 (2019).

[30] P. Chand, S. Vaish, P. Kumar, Physica B Condens. Matter.524, 53-63(2017). https://doi.org/10.1016/j.physb.2017.08.060

[31] K. Shetty, L. Renuka, H. Nagaswarupa, H. Nagabhushana, K. Anantharaju, D. Rangappa, S. Prashantha, K. Ashwini, Mater. Today: Proc.4, 11806-11815 (2017). https://doi.org/10.1016/j.matpr.2017.09.098

[32] M. Hashim, S. Kumar, B. Koo, S.E. Shirsath, E. Mohammed, J. Shah, R. Kotnala, H. Choi, H. Chung, R. Kumar, J. Alloys Compd. 518, 11-18 (2012). https://doi.org/10.1016/j.jallcom.2011.12.017

[33] K.S. Rao, S.R. Nayakulu, M.C. Varma, G. Choudary, K. Rao, J. Magne. Magne. Mater.451,602-608 (2018). https://doi.org/10.1016/j.jmmm.2017.11.069

[34] S. Feng, W. Yang, Z. Wang, Mater. Sci. Eng. B.176, 1509-1512 (2011). https://doi.org/10.1016/j.mseb.2011.09.007

[35] M. Dar, D. Varshney, J. Magne. Magne. Mater. 436, 101-112 (2017). https://doi.org/10.1016/j.jmmm.2017.04.046

[36] A. Džunuzović, N. Ilić, M.V. Petrović, J. Bobić, B. Stojadinović, Z. Dohčević-Mitrović, B. Stojanović, J. Magne. Magne. Mater.374, 245-251 (2015). https://doi.org/10.1016/j.jmmm.2014.08.047

[37] A. Manikandan, R. Sridhar, S.A. Antony, S. Ramakrishna, Mater. Sci. Eng. B1076, 188-200 (2014). https://doi.org/10.1016/j.molstruc.2014.07.054

[38] N. Ajinkya, X. Yu, P. Kaithal, H. Luo, P. Somani, S. Ramakrishna, Mater.13, 4644 (2020). https://doi.org/10.3390/ma13204644

[39] P. Dhiman, G. Rana, A. Kumar, G. Sharma, D.-V.N. Vo, T.S. AlGarni, M. Naushad, Z.A. ALOthman, Chem. Eng. Res. Des.175, 85-101 (2021). https://doi.org/10.1016/j.cherd.2021.08.028

[40] A. Salunkhe, V. Khot, M.R. Phadatare, N. Thorat, R. Joshi, H. Yadav, S. Pawar, J.Magne. Magne. Mater. 352, 91-98 (2014). https://doi.org/10.1016/j.jmmm.2013.09.020

[41] A. Alarifi, N. Deraz, S. J. Alloys Comp.486 501-506 (2009). https://doi.org/10.1016/j.jallcom.2009.06.192

[42] P. Dhiman, M. Patial, A. Kumar, M. Alam, M. Naushad, G. Sharma, D.-V.N. Vo, R. Kumar, Mater. Lett.284, 129005 (2021). https://doi.org/10.1016/j.matlet.2020.129005

[43] E. Hema, A. Manikandan, M. Gayathri, M. Durka, S.A. Antony, B. Venkatraman, J. Nanosci. Nanotechnol.16, 5929-5943 (2016). https://doi.org/10.1166/jnn.2016.11037

[44] A.B. Naik, P.P. Naik, S.S. Hasolkar, D. Naik, Ceram. Int.46, 21046-21055 (2020). https://doi.org/10.1016/j.ceramint.2020.05.177

[45] H. Kaur, A. Singh, V. Kumar, D.S. Ahlawat, J. Magne. Magne. Mater.474, 505-511 (2019). https://doi.org/10.1016/j.jmmm.2018.11.010

[46] A. Baykal, N. Kasapoğlu, Y. Köseoğlu, A.C. Başaran, H. Kavas, M.S. Toprak, Cent. Eur. J. Chem.6, 125-130 (2008).

[47] W. Zhong, W. Ding, Y. Jiang, N. Zhang, J. Zhang, Y. Du, Q. Yan, J. Am. Ceram. Soc.80, 3258-3262 (1997). https://doi.org/10.1111/j.1151-2916.1997.tb03264.x

[48] S.K. Mishra, L. Pathak, V. Rao, Mater. Lett.32, 137-141 (1997). https://doi.org/10.1016/S0167-577X(97)00027-X

[49] K. Martirosyan, E. Galstyan, S. Hossain, Y.-J. Wang, D. Litvinov, Mater. Sci. Eng. B176, 8-13 (2011). https://doi.org/10.1016/j.mseb.2010.08.005

[50] X. Huang, J. Zhang, L. Wang, Q. Zhang, J. alloys comp.540, 137-140 (2012). https://doi.org/10.1016/j.jallcom.2012.05.015

[51] P.A. Vinosha, A. Manikandan, A.S.J. Ceicilia, A. Dinesh, G.F. Nirmala, A.C. Preetha, Y. Slimani, M. Almessiere, A. Baykal, B. Xavier, Ceram.Inter.47, 10512-10535 (2021). https://doi.org/10.1016/j.ceramint.2020.12.289

[52] K. Huang, J. Yu, L. Zhang, J. Xu, P. Li, Z. Yang, C. Liu, W. Wang, X. Kan, J. Alloys Comp.825, 154072 (2020). https://doi.org/10.1016/j.jallcom.2020.154072

[53] W. Chen, W. Wu, C. Zhou, S. Zhou, M. Li, Y. Ning, J. Electron. Mater.47, 2110-2119 (2018). https://doi.org/10.1007/s11664-017-6021-8

[54] X. Wu, W. Chen, W. Wu, Y. Ning, S. Chen, J. MATER. SCI-MATER. EL.28, 18815-18824 (2017). https://doi.org/10.1007/s10854-017-7831-4

[55] C.H. Chia, S. Zakaria, M. Yusoff, S. Goh, C.Y. Haw, S. Ahmadi, N.M. Huang, H.N. Lim, Ceram. Inter. 36, 605-609 (2010). https://doi.org/10.1016/j.ceramint.2009.10.001

[56] M. Houshiar, F. Zebhi, Z.J. Razi, A. Alidoust, Z. Askari, J. Magne. Magne. Mater.371, 43-48 (2014). https://doi.org/10.1016/j.jmmm.2014.06.059

[57] M. Zahraei, A. Monshi, M. del Puerto Morales, D. Shahbazi-Gahrouei, M. Amirnasr, B. Behdadfar, J. Magne. Magne. Mater.393, 429-436 (2015). https://doi.org/10.1016/j.jmmm.2015.06.006

[58] R. Valenzuela, Phy. Res. Inte. (2012).

[59] F. Pereira, (Fortaleza,2009).

[60] I.-C. Tung, J.-S. Yang, T.-S.C.T.-S. Chin, Jpn. J. Appl. Phys.36, 5091 (1997). https://doi.org/10.1143/JJAP.36.5091

[61] D. Evans, G. Fischer, J. Geiger, F. Martin, J. AM. CERAM. SOC.63, 629-634 (1980). https://doi.org/10.1111/j.1151-2916.1980.tb09850.x

[62] K. Haneda, C. Miyakawa, H. Kojima, J. Amer. Ceram.Soc.57, 354-357 (1974). https://doi.org/10.1111/j.1151-2916.1974.tb10921.x

[63] H. Yamamoto, H. Kumehara, R. Takeuchi, H. Nishio, J. phys., IV7, C1-535-C531-536(1997).

[64] J. Wang, G. Huang, X. Zhong, L. Sun, Y. Zhou, E. Liu, Appl. Phys. Lett.88, 252502 (2006). https://doi.org/10.1063/1.2208564

[65] M. Rashad, I. Ibrahim, J. Magne. Magne. Mater.323, 2158-2164 (2011). https://doi.org/10.1016/j.jmmm.2011.03.023

[66] F. Pereira, A. Sombra, in: Solid State Phenom., Trans Tech Publ, 1-64 (2013). https://doi.org/10.4028/www.scientific.net/SSP.202.1

[67] D. Lisjak, M. Drofenik, J. Europ. Ceram. Soc.27, 4515-4520 (2007). https://doi.org/10.1016/j.jeurceramsoc.2007.02.202

[68] D. Lisjak, M. Drofenik, J. Europ. Ceram. Soc.26, 3681-3686 (2006). https://doi.org/10.1016/j.jeurceramsoc.2005.12.014

[69] S. Hu, J. Liu, H. Yu, Z. Liu, J. Magne. Magne. Mater.473, 79-84 (2019). https://doi.org/10.1016/j.jmmm.2018.10.044

[70] M. Sajjia, M. Oubaha, M. Hasanuzzaman, A. Olabi, Ceram. Inter.40, 1147-1154 (2014). https://doi.org/10.1016/j.ceramint.2013.06.116

[71] L. Wang, M. Lu, Y. Liu, J. Li, M. Liu, H. Li, Ceram. Inter.41, 4176-4181 (2015). https://doi.org/10.1016/j.ceramint.2014.12.099

[72] R. Zhang, L. Sun, Z. Wang, W. Hao, E. Cao, Y. Zhang, Mater. Res. Bull.98, 133-138 (2018). https://doi.org/10.1016/j.materresbull.2017.08.006

[73] M. Zate, S. Raut, S.D. Shirsat, S. Sangale, A. Kadam, (Elsevier, 2020).

[74] E. Matijevic, J. Colloid Interface Sci.117, 593-595 1987).
https://doi.org/10.1016/0021-9797(87)90426-7

[75] I.J. McColm, N. Clark, Blackie and Son Ltd., 1988, 345 (1988).

[76] C. Sürig, D. Bonnenberg, K. Hempel, P. Karduck, H. Klaar, C. Sauer, J. phys., IV7, C1-315-C311-316 (1997).

[77] R. Pullar, M. Stacey, M. Taylor, A. Bhattacharya, Act. mater.49, 4241-4250 (2001).
https://doi.org/10.1016/S1359-6454(01)00304-4

[78] R. Pullar, S. Appleton, A. Bhattacharya, J. Magne. Magne. Mater.186, 326-332 (1998). https://doi.org/10.1016/S0304-8853(98)00107-3

[79] R. Pullar, A. Bhattacharya, Mater. Lett.57, 537-542 (2002).
https://doi.org/10.1016/S0167-577X(02)00825-X

[80] N. Yasmin, M.Z. Iqbal, M. Zahid, S.F. Gillani, M.N. Ashiq, I. Inam, S. Abdulsatar, M. Safdar, M. Mirza, Ceram. Inter.45, 462-467 (2019).
https://doi.org/10.1016/j.ceramint.2018.09.190

[81] I. Auwal, A. Baykal, S. Güner, H. Sözeri, Ceram. Inter.43, 1298-1303 (2017).
https://doi.org/10.1016/j.ceramint.2016.10.080

[82] D. Kotsikau, M. Ivanovskaya, V. Pankov, Y. Fedotova, Solid State Sci.39, 69-73 (2015). https://doi.org/10.1016/j.solidstatesciences.2014.11.013

[83] S. Sarıtaş, B.C. Şakar, E. Turgut, M. Kundakci, M. Yıldırım, Mater. Today: Proc.46, 7025-7029 (2021). https://doi.org/10.1016/j.matpr.2021.03.284

[84] Z. Tang, S. Nafis, C. Sorensen, G. Hadjipanayis, K. Klabunde, IEEE Trans. Magn.25, 4236-4238 (1989). https://doi.org/10.1109/20.42580

[85] C. Barathiraja, A. Manikandan, A. Uduman Mohideen, S. Jayasree, S.A. Antony, J. Supercond. Nov. Magn.29, 477-486 (2016). https://doi.org/10.1007/s10948-015-3312-2

[86] P. Thakur, D. Chahar, S. Taneja, N. Bhalla, A. Thakur, Ceram. inter.46, 15740-15763 (2020). https://doi.org/10.1016/j.ceramint.2020.03.287

[87] M. Sundararajan, L.J. Kennedy, U. Aruldoss, S.K. Pasha, J.J. Vijaya, S. Dunn, Mater. Sci. Semicond Process40, 1-10 (2015).
https://doi.org/10.1016/j.mssp.2015.06.002

[88] B. Grindi, Z. Beji, G. Viau, A. BenAli, J. Magne. Magne. Mater.449, 119-126(2018). https://doi.org/10.1016/j.jmmm.2017.10.002

[89] K.K. Senapati, C. Borgohain, P. Phukan, J. Mol. Catal. A Chem.339, 24-31 (2011). https://doi.org/10.1016/j.molcata.2011.02.007

[90] V. Pillai, P. Kumar, M. Hou, P. Ayyub, D. Shah, Adv. Colloid Interface Sci.55, 241-269 (1995). https://doi.org/10.1016/0001-8686(94)00227-4

[91] D. Rawlinson, P. Sermon, J. phys., IV7, C1-755-C751-756 (1997).

[92] L.G. Cerda, S.M. Montemayor, J. Magne. Magne. Mater.294, e43-e46 (2005). https://doi.org/10.1016/j.jmmm.2005.03.051

[93] P.R. Varma, R.S. Manna, D. Banerjee, M.R. Varma, K. Suresh, A. Nigam, J. Alloys Comp.453, 298-303 (2008). https://doi.org/10.1016/j.jallcom.2006.11.058

[94] V. Dutta, S. Sharma, P. Raizada, A. Hosseini-Bandegharaei, V.K. Gupta, P. Singh, J. Saudi Chem. Soc. 23, 1119-1136 (2019). https://doi.org/10.1016/j.jscs.2019.07.003

[95] R. Sridhar, R. Dachepalli, K. K Vijaya, Adv. Mater. Chem. Phys. (2012).

[96] V. Sankaranarayanan, Q. Pankhurst, D. Dickson, C. Johnson, J. Magne. Magne. Mater. 125, 199-208 (1993). https://doi.org/10.1016/0304-8853(93)90838-S

[97] Y. Li, Q. Wang, H. Yang, Curr. Appl. Phys.9, 1375-1380 (2009). https://doi.org/10.1016/j.cap.2009.03.002

[98] L. Junliang, Z. Wei, G. Cuijing, Z. Yanwei, J.Alloys Comp.479, 863-869 (2009). https://doi.org/10.1016/j.jallcom.2009.01.081

[99] J. Huang, H. Zhuang, W. Li, Mater. Res. Bull.38, 149-159 (2003). https://doi.org/10.1016/S0025-5408(02)00979-0

[100] S.S. Kumar, R.K. Singh, N. Kumar, G. Kumar, U. Shankar, Mater. Today: Proc.46, 8567-8572 (2021). https://doi.org/10.1016/j.matpr.2021.03.547

[101] A.-H. El Foulani, A. Aamouche, F. Mohseni, J. Amaral, D. Tobaldi, R. J.Alloys Comp.774, 1250-1259 (2019). https://doi.org/10.1016/j.jallcom.2018.09.393

[102] M.G. Naseri, E.B. Saion, H.A. Ahangar, M. Hashim, A.H. Shaari, J. Magne. Magne. Mater.323, 1745-1749 (2011). https://doi.org/10.1016/j.jmmm.2011.01.016

[103] A. Ataie, M. Piramoon, I. Harris, C. Ponton, J. mater. sci.30, 5600-5606 (1995). https://doi.org/10.1007/BF00356692

[104] M. Jean, V. Nachbaur, J. Bran, J.-M. Le Breton, J. Alloys Comp.496,306-312 (2010). https://doi.org/10.1016/j.jallcom.2010.02.002

[105] S. Shafiu, H. Sözeri, A. Baykal, J. Supercond. Nov. Magn.27, 1593-1598 (2014). https://doi.org/10.1007/s10948-014-2490-7

[106] J.-H. Lee, T.-B. Byeon, H.-J. Lee, C.-G. Kim, T.-O. Kim, J. Phys. IV, 7, C1-751-C751-752 (1997).

[107] X. Liu, J. Wang, L.-M. Gan, S.-C. Ng, J. Magne. Magne. Mater.195, 452-459 (1999). https://doi.org/10.1016/S0304-8853(99)00123-7

[108] S.A. Morrison, C.L. Cahill, E.E. Carpenter, S. Calvin, R. Swaminathan, M.E. McHenry, V.G. Harris, J. Appl. Phys.95, 6392-6395 (2004). https://doi.org/10.1063/1.1715132

[109] C. Singh, S. Jauhar, V. Kumar, J. Singh, S. Singhal, Mater.Chem. Phys.156, 188-197 (2015). https://doi.org/10.1016/j.matchemphys.2015.02.046

[110] S. Shanmugam, B. Subramanian, Mater. Sci. Eng. : B 252, 114451 (2020). https://doi.org/10.1016/j.mseb.2019.114451

[111] A.M. Ibrahim, M. Abd El-Latif, M.M. Mahmoud, J. Alloys Comp.506, 201-204 (2010). https://doi.org/10.1016/j.jallcom.2010.06.177

[112] V. Chaudhary, A.K. Kaushik, H. Furukawa, A. Khosla, ECS Sensors Plus, (2022).

[113] M.K. Enamala, M. Chavali, A. Tangellapally, D. Pasumarthy, M.K. Murthy, C. Kuppam, V. Chaudhary, R. Mishra, D. Naradasu, (Elsevier, 2021) 243-259. https://doi.org/10.1016/B978-0-12-820498-6.00010-X

[114] G. Kogias, V. Tsakaloudi, P. Van der Valk, V. Zaspalis, J. Magne. Magne. Mater.324, 235-241 (2012). https://doi.org/10.1016/j.jmmm.2011.07.055

[115] M. Abbas, B.P. Rao, M.N. Islam, K.W. Kim, S. Naga, M. Takahashi, C. Kim, Ceram. inter.40, 3269-3276 (2014). https://doi.org/10.1016/j.ceramint.2013.09.109

[116] W. Zhong, W. Ding, N. Zhang, J. Hong, Q. Yan, Y. Du, J. Magne. Magne. Mater. 168,196-202 (1997). https://doi.org/10.1016/S0304-8853(96)00664-6

[117] G. Mendoza-Suárez, M. Cisneros-Morales, M. Cisneros-Guerrero, K. Johal, H. Mancha-Molinar, O. Ayala-Valenzuela, J. Escalante-Garcıa, Mater. chem. phys.77, 796-801 (2003). https://doi.org/10.1016/S0254-0584(02)00141-4

[118] H.-F. Yu, K.-C. Huang, J. Magne. Magne. Mater.260, 455-461 (2003). https://doi.org/10.1016/S0304-8853(02)01389-6

[119] D. Mishra, S. Anand, R. Panda, R. Das, Mater.chem. phys.86, 132-136 (2004). https://doi.org/10.1016/j.matchemphys.2004.02.017

[120] D. Lisjak, K. Bobzin, K. Richardt, M. Bégard, G. Bolelli, L. Lusvarghi, A. Hujanen, P. Lintunen, M. Pasquale, E. Olivetti, J. Europ. Ceram. Soc.29, 2333-2341 (2009). https://doi.org/10.1016/j.jeurceramsoc.2009.01.028

An Introduction to Hard Ferrites: From Fundamentals to Practical Applications   Materials Research Forum LLC
Materials Research Foundations **142** (2023) 66-92   https://doi.org/10.21741/9781644902318-3

Chapter 3

Effect of Substitution on the Dielectric and Magnetic Properties of $BaFe_{12}O_{19}$

Hitanshu Kumar[1,*], Arashdeep Singh[2], Kavita Rana[3]

[1]Department of Applied Sciences, Modern Group of Colleges, Mukerian, Punjab, India

[2]Department of Mechanical Engineering, Modern Group of Colleges, Mukerian, Punjab, India

[3]Department of Allied Health Sciences, Modern Group of Colleges, Mukerian, Punjab, India

*hitanshuminhas@gmail.com

Abstract

Hexaferrites are composed of irons and many other divalent metal ions in various atomic ratios. They are classified as M, W, X, Y, Z, and U type according to the crystalline structure. They are also used as ferromagnetic compounds. These materials have been utilized as the permanent magnets, magnetic recording media and microwave absorbers due to their superior properties. These materials also have wide application in ferrofluids, sensors, loudspeakers, electric power generation, rotors in small direct current (DC) motors, automotive electronics, ferrite cores, fabrication of inductors, micro electro mechanical systems (MEMS), capacitors, transistors, microwave, and in magneto static and electromagnetic devices. $BaFe_{12}O_{19}$ has gain focus in recent years due to its large coercivity, high magneto crystalline anisotropy, relatively large saturation magnetization, electrical resistivity, low cost, and ability to resist corrosion. The Present chapter will focus on the effect of substitution on the dielectric and magnetic properties $BaFe_{12}O_{19}$. Morphology and properties of barium hexaferrites, effect of Co-Ti substitution on magnetic properties of nanocrystalline $BaFe_{12}O_{19}$ and effect of rare-earth materials substitution on the micro structural and magnetic properties of $BaFe_{12}O_{19}$ are covered.

Keywords

Hexaferites, Dielectric, Ferrofluids, Barium, Nanocrystalline

Contents

1.   Introduction

Since their discovery more than half a century ago, hexagonal ferrites have received tremendous attention as commercially and technologically important materials. Hexagonal ferrites also recognized as 'hexaferrites' exclusively report for more than 50% of the total magnetic materials produced worldwide owing to their applications in magnetic recording, ferrofluids, sensors, loudspeakers, electric power generation, rotors in small direct current (DC) motors, automotive electronics, ferrite cores, fabrication of inductors, micro electro mechanical systems (MEMS), capacitors, transistors, microwave, and in magneto static and electromagnetic devices [1–4]. Along with the family of hexaferrites, which includes M, Z, Y, W, X and U-type ferrites, M-Type hexaferrites, especially BaFe$_{12}$O$_{19}$ has caught the consideration of researchers in recent years owing to their extraordinary properties including large coercivity, high magneto crystalline anisotropy, relatively large saturation magnetization, electrical resistivity, low cost, ability to resist corrosion, low eddy current, high Curie temperature and low dielectric losses [4–8]. BaFe$_{12}$O$_{19}$ has a space group P63/mmc and possesses hexagonal structure resembling that of magneto plumbite, which is a naturally occurring mineral. A single molecular unit of M-type barium hexaferrites (BaM) comprises four alternating blocks of hexagonally packed (S) and cubically packed (R) layers in a series.

Magnetic applications of BaM can be increased by modifying/improving its magnetic properties including saturation magnetization, remnant magnetization and coercivity, which can be done by substituting Fe^{3+} ions at the crystallographic positions. The literature survey shows that researchers have substituted Fe^{3+} ions with other trivalent ions such as Al^{3+} [2], Ce^{3+}[5], Sm^{3+} [6], or in combination with divalent and tetravalent ions like Mg^{2+}/Ti^{2+}[7], Ni^{+2}/Sn^{+4} [10], Ni^{+2}/Ti^{+4} [10], Ni^{+2}/Zr^{+4} [11] etc. For substitution of a divalent impurity such as Ni^{+2}, Co^{+2}, etc., charge is balanced by vacancies forming in the structure, which may occur at any crystallographic position of Fe^{3+} ions causing modification/improvement of magnetic properties [12]. Ni is a good candidate for the substitution of Fe in barium hexaferrites due to similar ionic radii and electronic configuration of Ni and Fe. The effect of Ni^{2+} ions on the properties of $BaFe_{12}O_{19}$ has been studied in combination with various tetravalent impurities [10-11, 13] which resulted in the enhancement of magnetic and microwave absorbing properties of $BaFe_{12}O_{19}$. In single crystals, addition of Ni^{2+} caused a decrease in saturation magnetization Ms and coercivity Hc [12-13]. However, little work has been done on the substitution of solely Ni^{2+} in polycrystalline $BaFe_{12}O_{19}$ hexagonal ferrites, which suggests the obvious need of studying the role of Ni^{2+} ions in Ni co-doped $BaFe_{12}O_{19}$. The effect of Nickel (as a dopant) on the structure, microstructure, magnetic and electrical properties of $BaFe_{12}O_{19}$ has been investigated.

1.1   Magnetic properties and morphology of copper-substituted barium hexaferrites

Hexaferrites, composed of iron ions and other divalent metal ions in various atomic ratios, can be classified according to the crystalline structure as M, W, X, Y, Z, and U types. Graphical presentation of these structures can be found in an excellent review [14]. The M-type hexaferrites, especially barium hexaferrite ($BaFe_{12}O_{19}$), have been intensively studied and implemented as permanent magnets due to their high intrinsic coercivity (Hc), large crystalline anisotropy, high chemical stability, and low cost. Properties of barium hexaferrites at high frequencies were implemented in microwave absorbers [15, 16]. Their incorporations into polymer composites were investigated to improve forming and mechanical properties [17, 18]. There is growing interest in hard/soft magnetic composites with tunable properties, including the combination of barium hexaferrites with magnesium ferrites [19]. Furthermore, potential applications of barium hexaferrites in battery cathodes and magnetic fluids have been proposed [20, 21]. Since barium hexaferrite has already been one of the most used magnetic materials with sizeable global market values, its alterable magnetic properties are increasingly being investigated. In a magnetoplumbite structure of barium hexaferrites, Fe ions are located on five different crystallographic sites, and due to the interactions with O_2 ions, the diverse magnetic properties can be obtained

[14]. In addition, the structural substitution by various transition-metal ions modifies magnetic hysteresis loops. Examples are Cu^{2+} [22-26], Co^{2+} [27], Sm^{3+} [28], Ga^{3+} [29], Cr^{3+} [30], and Ce^{3+} [31]. Moreover, the substitutions of Ce-Co [15], Sm-Co [32], Cu-Zr [33], Co-Zr [34], and La-Mn [35] have been combined in M-type hexaferrites. However, because the magnetic and dielectric properties of ferrites depend on their microstructures and crystal structures depicted in [14, 36-37], the different values of coercivity and magnetization have been reported. Of particular relevance to this report is the partial substitution in the barium hexaferrite structure by Cu^{2+}, which can be either $Ba_{1-x}Cu_xFe_{12}O_{19}$ or $BaFe_{12-x}Cu_xO_{19}$.

Asiri et al. synthesized $Ba_{1-x}Cu_xFe_{12}O_{19}$ using the citrate sol-gel combustion method. The coercivity was substantially decreased to 1726 Oe in the case of x = 0.1 but increased from 2121 to 2460 Oe with increasing x from 0.2 to 0.4. By contrast, the saturation magnetization was increased to 54.36 emu/g in the case of x = 0.1 but reduced by higher substitution levels [38]. According to the AC susceptibility measurement by Slimani et al. [39], the substitution of Ba^{2+} by Cu^{2+} strongly affected the blocking temperature of $Ba_{1-x}Cu_xFe_{12}O_{19}$. For $Ba_{1-x}Cu_xFe_{12}O_{19}$, the citrate sol-gel combustion method was also used to study the higher x up to 2 [40]. Kumar et al. reported the maximum magnetization and the lowest coercivity in the case of x = 1. The magnetic as well as dielectric properties were correlated with lattice parameters [40]. Alternatively, Rafiq et al. employed the solid-state mixed oxide method. Low coercivities of 932.5 and 262.1 Oe were, respectively, obtained in $BaFe_{11.9}Cu0._1O_{19}$, and $BaFe_{11.7}Cu_{0.3}O_{19}$. Interestingly, the coercivity was increased to 1911 Oe with a further increase of Cu to x = 0.5 [41]. The research works demonstrate that Cu substitution is a promising route to the commercial production of barium hexaferrites. Variations in sites and levels of partial substitution lead to the coercivity and magnetization suitable for various applications ranging from conventional permanent magnets, recording media, microwave absorbers for novel magnetic fluids, and electrodes. However, the results indicate that the effect of Cu on the magnetic properties of the barium hexaferrite needs more investigation. From various methods developed to synthesize and dope ferrites [42], the sol-gel auto combustion method was selected in this research work to synthesize $Ba_{1-x}Cu_xFe_{12}O_{19}$ (x = 0, 0.1, 0.3, and 0.5) for its cost-effectiveness [43]. Magnetic properties of $BaFe_{12-x}Cu_xO_{19}$ with Cu substitution up to x = 0.5 were compared and correlated to their phase and morphology. The key effects could then be identified and controlled during the synthesis.

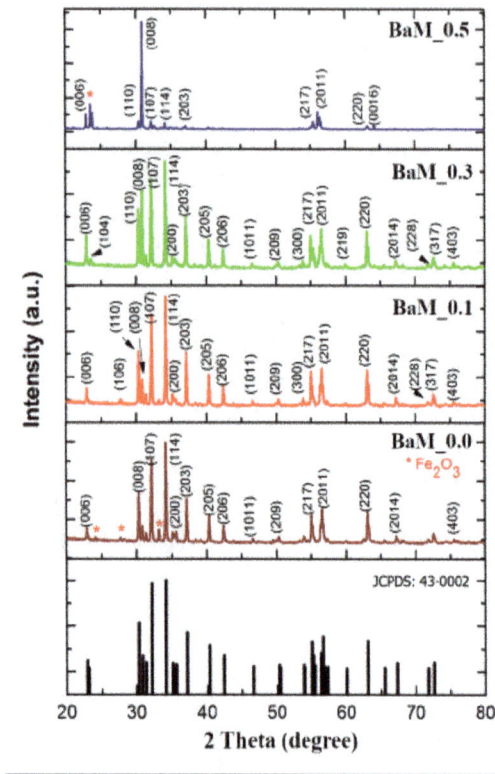

Figure 1. XRD spectra of $BaFe_{12-x}Cu_xO_{19}$ (x = 0, 0.1, 0.3, and 0.5) compared to a standard profile of barium hexaferrite (JCPDS: 43-0002)

The copper (Cu) substitution in barium hexaferrite ($BaFe_{12}O_{19}$) crystals from the sol-gel auto-combustion synthesis is demonstrated as a cost-effective pathway to achieve alterable magnetic properties. Subsequent heat treatments at 450°C and 1050°C result in irregularly shaped nanoparticles characterized as the M-type $BaFe_{12}O_{19}$ with the secondary phase of hematite (αFe_2O_3). Despite the mixed phase, the substantial coercivity of and magnetization as high as 74.8 emu/g are obtained in this undoped ferrite. The majority of particles become micro rods for x = 0.1 and micro plates in the case of x = 0.3 and 0.5. The coercivity and magnetization tend to reduce as Cu^{2+} increasingly substitutes Fe^{3+}. From these findings, magnetic properties for various applications in microwave absorbers, recording media, electrodes, and permanent magnets can be tailored by the partial substitution in hexaferrite crystals. The c/a ratio and the V_{cell} are not sensitive to the variation in Cu substitution from 0.0–0.5. The respective values around 3.94 and 699 (Å)^3 are slightly higher than those reported in previous experiments [38, 41].

Table 1. *Comparison of structural parameters from XRD, remanent magnetization, and magnetic squareness from VSM of $BaFe_{12-x}Cu_xO_{19}$ samples.*

Sample	C/a Ratio	Vcell	Crystallite Size (nm)	Remanent magnetization (emu/g)	Magnetic Squareness
BaM_0.0 (x= 0.0)	3.94	699	71	35.8 ± 0.8	0.479
BaM_0.1 (x= 0.1)	3.94	699	68	22.5 ± 1.9	0.402
BaM_0.3 (x= 0.3)	3.94	699	72	25.0 ± 1.1	0.379
BaM_0.5 (x= 0.5)	3.94	697	96	7.2 ± 3.5	0.132

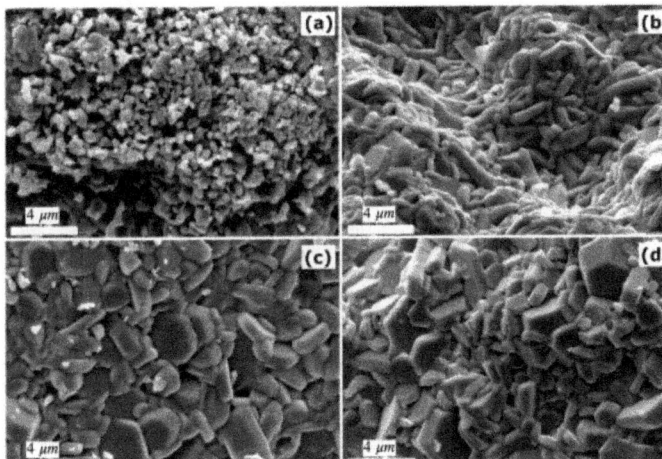

Figure 2. *SEM micrographs of BaFe12−xCuxO19; (a) x = 0, (b) x = 0.1, (c) x = 0.3, and (d) x = 0.5*

Muhammad Asif Rafiq et al. represent very nice results on this topic i.e. average grain sizes are presented in Table 2. From a closer look at the SEM images, large voids among the grains can easily be observed, which might be present due to the oxygen vacancies produced during synthesis at higher temperatures (1300°C). Oxygen vacancies are the most general defects, which can easily diffuse and might have played an important role in the grain growth [44]. Morphology of particles is platelet-like for all samples, which is an expected result, keeping in view the anisotropy present in barium hexaferrites as also indicated by the lattice parameters calculated from XRD profiles. With the addition of Ni, an increase in the coercivity of $BaFe_{12}O_{19}$ was observed as can be seen in Table 2. It is well known about barium hexaferrites that their coercivity is related to the uniaxial magneto crystalline anisotropy along the c axis, and the increase in coercivity (Hc) in this case can be attributed to the enhancement of magneto crystalline anisotropy by the substitution of Ni^{2+} [45]. Introduction of defects in ceramics is inevitable owing to the fact that they are processed at very high temperatures. It is also known that defects like pores, holes and oxygen vacancies can hinder the movement of magnetic domains and are responsible for the increase in the coercivity. In the present work, increase in defects, especially oxygen vacancies are expected due to the difference in the valency of the substituents, i.e., Fe and Ni which has already been observed. This can explain the increase in coercivity and a similar kind of behavior has already been reported for other ceramic systems like $BiFeO_3$ and Potassium-Sodium Niobates [44, 46].

Table 2. Magnetic properties and average grain size of $BaFe_{12-x}Ni_xO_{19}$ for x =0, 0.3 and 0.5

X	Saturation magnetization Ms (emu/g)	Remanent magnetization Mr (emu/g)	Coercivity He (Oe)	Average Size (μm)
0	55.35	20.24	1027.2	0.9723
0.3	57.57	20.65	1240.6	1.4147
0.5	56.42	20.51	1971.6	1.9005

1.2   Effect of Co-Ti substitution on magnetic properties of nanocrystalline $BaFe_{12}O_{19}$

Recently, investigations of barium ferrite $BaFe_{12}O_{19}$ as hard magnetic material have been developed. The enhancement of magnetic properties such as saturated magnetization (Ms), remanence (Mr), and coercive force (Hc) become major interest to be investigated. An effect of the various processes [47] and designing crystal structure of the barium ferrite to magnetic properties were investigated systematically [48-51]. In order to improve the magnetic properties, the nanocrystalline $BaFe_{12}O_{19}$ ferrites have potential opportunity to be developed. As reported, the development of the barium ferrite compositions was substituted by Ba and (Sr) [52-54]. In order to understand the magnetic properties of $BaFe_{12}O_{19}$, the Fe was substituted by Al, Sn, NiSn, Ti, or Co [55-59]. Furthermore, the improvement of coercivity of $BaFe_{12}O_{19}$ is needed to realize magnetic absorption materials microwave [60-62]. In this report, nanocrystalline $BaFe_{12-2x}Co_xTi_xO_{19}$ with x = 0, 1, 2, 3 were synthesized. The formations of polycrystalline samples that the cationic of Co^{2+} and Ti^{4+} in Co-Ti substituted Fe in $BaFe_{12}O_{19}$ ferrite were prepared by solid state reaction method. The crystal structure, microstructure and magnetic properties of $BaFe_{12-x}Co_xTi_xO_{19}$ will be discussed systematically.

Fig. 2 shows the hysteresis loops of the magnetic behavior of the as-synthesized barium ferrite. Meanwhile, the Table 3 is the data for hysteresis loops of the barium ferrite. The undoped barium ferrite samples have larger coercive force (Hc), area, and the higher remanence (Mr) than doped barium ferrite. The coercive force of barium ferrite (x=0) is about 85.06 kA/m. While the doped barium ferrites rapidly decrease up to 0.939 kA/m at x=1. The reduction of the crystal anisotropy in x=1 cause the weak uni-axial anisotropy along the c-axis of the doped barium ferrite. Substitution of Co-Ti decreases the coercive force of barium ferrite significantly. Further, Co-Ti increases the coercivity Hc= 24.028 kA/m at x=2 and Hc= 23,987 kA/m at x=3. These results were predicted due to increasing the crystal anisotropy. Furthermore, the low coercivity explains that the doped barium ferrite is the soft magnetic material.

Table 3. Data for hysteresis loops of the barium ferrite.

Composition	Mr (T)	Hc (kA/m)	Ms (T)	Mr/Ms (T)
0	0.045	85.06	0.124	0.365
1	0.02	0.939	0.084	0.234
2	0.004	24,028	0.024	0.159
3	0.005	23,987	0.015	0.372

The formation of polycrystalline samples that the cationic of Co^{2+} and Ti^{4+} in Co-Ti substituted Fe in $BaFe_{12}O_{19}$ ferrites structure were ready by solid state reaction method. The results show that the nanocrystalline $BaFe_{12-2x}Co_xTi_xO_{19}$ has single phase with polycrystalline structure, the grain size decrease by substitution, and the coercivity (Hc) and saturation magnetization (Ms) decrease with increasing Co-Ti substitutions.

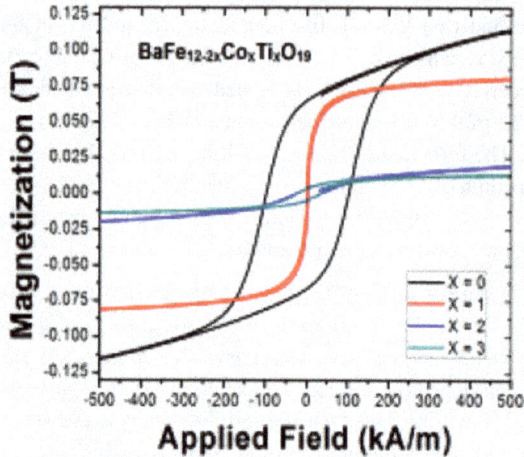

Figure 3. The room temperature hysteresis loop of $BaFe_{12-2x}Co_xTi_xO_{19}$

1.3   Effect of rare-earth materials substitution on the micro structural and magnetic properties of $BaFe_{12}O_{19}$

Hexaferrites with the basic $MFe_{12}O_{19}$ (M = Ba, Sr) formula are doped and they are frequently used as ferrimagnetic compounds. These materials have been utilized as the permanent magnets, magnetic recording media and microwave absorbers due to their superior properties [63-65]. That is why, the attempts to improve the magnetic properties of the hexaferrites are still among the recent research topics. In order to achieve an improvement in the magnetic properties, the M element is replaced with a couple of rare-earth ions. At the room temperature, undoped barium hexaferrite exhibits a saturation magnetization of 70 emu/g [66-67]. The replacement of Ba and/or Sr crystallographic sites in the hexaferrite lattice by the rare earth cations results in changing the valence state of Fe^{3+} to Fe^{2+} which is located on the 2a site. The presence of Fe^{2+} cation provides the hexaferrite with a considerable rise in the magneto crystalline anisotropy property. Thus, these consciously doped hexaferrites have been reported to hold an improved coercivity (Hc), saturation magnetization (Ms) and other related magnetic properties [68-70]. Most of the recent hexaferrite researches are focused on Sr-hexaferrites. Single substitution for hexaferrites has been carried out in some researches where M-type hexagonal ferrites have

74

been doped by Sm^{3+} and Sr hexaferrite has been doped by Nd^{3+} cations [71-72]. In a couple of recent researches, Gd^{3+} and Tb^{3+} have been utilized as dopants for Sr-hexaferrites [73-74]. Nano-hexaferrites have been prepared using La-Nd, Mn-Y, Zr-Nd and Sm-Er co-dopants [75-78]. A comparative study has been shown in Table 4.

Table 4. Comparison of the average crystallite sizes of hexaferrites.

Synthesis method	Compound	Dp (nm)	References
sol-gel	BHF/CF nano-e	71-91	[46]
Conventional Ceramic	BHF/TiO2 nano-e	83	[47]
sol-gel auto-com.	BHF/Ti	42	[48]
Conventional Ceramic	BHF nano-p	49.1	[49]
self-assembly	BHF submicron-s	500	[50]
Conventional Ceramic	BHF	50-150	[7]

$CF:CoFe_2O_4$: nano-e: nanocomposite: sol-gel auto-com.: sol-gel auto-combustion : nano-p:nanoparticles submicron-s:submicron spheres: con. Ceramic: conventional ceramic

The co-substitution for the magnetization improvement of Sr-hexaferrites has been performed by Al-Ce, Nb-Y, Ce-Zn, Ce-Mn, La-Co and Dy-Al and in these studies sol-gel, co-precipitation and low temperature sintering methods have been carried out [79-85]. On the other hand, single La^{3+} and Nd^{3+} substitution has been done for Ba-hexaferrites. In addition, co-doped Tm-Tb, Eu-Nd, Sm-Co Ba-hexaferrites have also been investigated and have been shown that co-substitution of the rare-earth ions provides hexaferrites with a rise in the magnetization and/or coercivity [86-89]. Among other magnetic properties, some attempts have also been made to improve the maximum energy product, $(BH)_{max}$ for Sr hexaferrites and Ba-hexaferrites using La-Sm [90-91]. Researchers have different manufacturing methods to produce hexaferrites such as traditional ceramic methods [92-96] or several chemical methods [97-103]. To the best of our knowledge, there are no recent studies which have reported the simultaneous effect of the sintering temperature and La^{3+} -Y^{3+} dopants on the properties of barium hexaferrites. Thus, enhancing the microstructural and the ferrimagnetic properties of $BaFe_{12}O_{19}$ by rare earth co-substitution.

Table 5. Magnetization and coercivity of La-Y co-doped hexaferrites.

Sample	Magnetization (emu/g)		Coercivity (kOe)	Reference
	Ms	Mr		
BHF	54.02	24.51	3.10	[45]
SBHF	80.30	50.00	3.64	[54]
BHF	55.00	35.00	8.73	[55]
BHF composites	44.94	27.82	3.58	[56]
BHF	51.81	34.55	1.73	[28]
SHF	70.07	41.78	4.30	[57]

$Ba_{0.85}(La, Y)_{0.15}Fe_{12}O_{19}$ hexaferrite magnets were produced using the powder metallurgy method. The phase analysis of the ferrite magnets was carried out by X-ray diffraction (XRD) technique.

Magnetization and coercivity of La-Y co-doped hexaferrites has been shown in Table 5. Saturation magnetizations, Ms were determined to be 48.60 emu/g and 52.95 emu/g for the samples sintered at 1150 °C and 1250 °C, respectively whereas the remanent magnetizations, Mr were 29.26 emu/g and 31.17 emu/g. The coercivity, Hc decreased from 3.95 kOe to the value of 2.44 kOe with the sintering temperature due to the increase of the crystallite size. The squareness ratios (Mr/Ms) of the ferrimagnetic samples were different because the uniaxial anisotropies altered after sintering at 1150 °C and 1250 °C. The maximum energy product, (BH)$_{max}$ dropped from 35.81 kJ/m3 to 27.38 kJ/m3 when the sintering temperature increased. This result can be attributed to a mixture of higher magnetization and the lower coercivity.

1.4 The effect of Nb substitution on magnetic properties of $BaFe_{12}O_{19}$ nano hexaferrites

Barium M-type hexaferrites are extensively utilized due to their outstanding features such as a complex magneto-plumbite structure, higher saturation, higher coercivity, better dielectric properties [104-105]. This feature leads for many applications in isolators, phase shifters and circulators that can be used as apparatuses in military and commercial devises [106,107]. Recently, many researchers have been investigated the impact of rare-earth (RE) substitutions in Fe or Ba (Sr) sites on saturation magnetization, coercivity and anisotropy of M-type hexagonal ferrites [108-111]. Use a suitable ratio of rare earth ions, a significant change in both structural and magnetic properties will be obtained due to the influence of the rare earth ions on magneto crystalline anisotropy, coercive field with magnetization and on the grain growth [112].

Microstructural and magnetic properties of $BaFe_{12-x}Nb_xO_{19}$ ($0.0 \leq x \leq 0.1$) nano hexaferrites (NHFs) have been investigated. The properties of the $BaFe_{12-x}Nb_xO_{19}$ in different samples are confirmed by XRD powder patterns, scanning electron microscopy and Fourier Transform Infrared spectroscopy. The analysis of hysteresis loops divulged that the different products display hard ferromagnetic behavior at different temperatures. The deduced values of saturation magnetization (Ms), remanence (Mr), magneton number (nB) and magneto crystalline anisotropy constant (K_{eff}) are reduced for lower Nb content and then increased with further increasing the Nb content, reaching a maximum values for $BaFe_{11.9}Nb_{0.1}O_{19}$ (x=0.1) nano hexaferrite. The coercivity (Hc) and intrinsic coercivity (H_{ci}) are diminished for lower x and are comparable to that of pristine one for higher x. The squareness ratio (Mr/Ms) are fluctuating between 0.50 and 0.55, implying the uniaxial anisotropy for different $BaFe_{12-x}Nb_xO_{19}$ nano hexaferrite shown in table 6.

Table 6. Value of n_B, α, K_{eff} and H_{ei} Of $BaFe_{12-X}Nb_XO_{19}$ ($0.0 \leq X \leq 0.1$) nano hexaferrites.

X	n_B (μ_B)		A ($\times 10^6$ Oe^2)		K_{eff} ($\times 10^5$ Erg/g)	
	300K	10K	300K	10K	300K	10K
0.00	10.19	14.45	9.08	9.32	2.99	4.29
0.02	5.86	8.16	9.40	9.90	1.75	2.50
0.04	8.96	14.54	8.37	9.66	2.53	4.41
0.06	10.41	14.76	9.85	9.75	3.19	4.50
0.08	10.93	15.07	9.98	9.70	3.37	4.58
0.10	12.58	17.58	10.00	9.46	3.89	5.28

The variation in Ms could be also clarified by means of the magnetic moment (nB) of the produced products. The values of nB of $BaFe_{12-x}Nb_xO_{19}$ ($0.0 \leq x \leq 0.1$) NHFs are shown in Table 6.

The non-substituted sample exhibits values of nB equal to 10.19 and 14.45 μB at 300 K and 10 K, respectively. Evidently, the $BaFe_{11.98}Nb_{0.02}O_{19}$ (x=0.02) sample that has the lowest Ms, has the minimum nB with values equal to 5.86 and 8.16 μB at 300 and 10 K, respectively. However, the sample with x=0.1, which exhibits the highest Ms, has the maximum nB with values of 12.58 and 17.58 μB at 300 K and 10 K, respectively. The increase (or decrease) of nB implies the strengthening (or weakening) of the super-exchange interactions among the octahedral and tetrahedral sites in ferrites. It is very clear that the variations in Mr as a function of Nb content display similar variations in Ms at both 300 K and 10 K. Compared to the pristine sample, Mr decreases for lower Nb amount x=0.02 and then increases with increasing x. Mr is lowest for x=0.02 and highest for x=0.10. In previous published research reports, it was stated that the variations in Mr depend largely on the variations in Ms and on the net alignment of grains magnetization resulted from the exchange interactions among nanoparticles [23-24]. The estimated values of K_{eff} at different temperatures are presented in Table 6. For the pristine product, $Keff$ is around 2.99×105 and 4.29×105 Erg/g at 300 and 10 K, respectively. $Keff$ decreases for lower concentration (x=0.02) and then starts to increase with further increasing x, reaching maximum values for x=0.1. The increment in $Keff$ is frequently a result of the intensification of magnetic anisotropy sources and the strengthening of magnetic interactions amongst nanoparticles owing to the substitution [30-32]. This could also explain the variations in Ms and Mr. Furthermore, it is reported also that the squareness ratio (SQR = Mr/Ms) takes values of 0.50 for uniaxial anisotropy and of 0.83 for cubic anisotropy according to S-W approach [1-2]. The variations of SQR as a function of Nb content at RT and 10 K for various products are shown in Fig. 3. Evidently, the values of SQR are ranging between 0.50 and 0.55 at both 300 and 10 K, which indicate the uniaxial anisotropy for different produced nano hexaferrites.

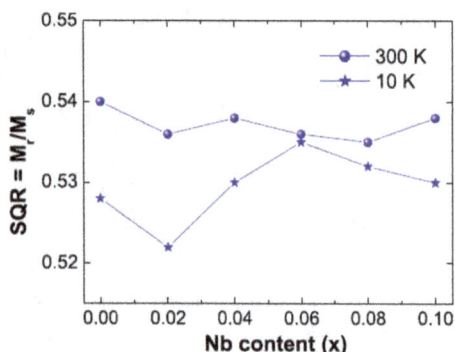

Figure 4. Variation of SQR for BAFE12-xNbxO19 (0.0≤x≤0.1) nano hexaferrites with respect to Nb content (x) at RT and 10 K.

1.5   Magnetic properties of Cu and Al doped nano BaFe$_{12}$O$_{19}$

M-type hexaferrite, the composition of which is represented by MeFe$_{12}$O$_{19}$ (Me = Ba, Sr or Pb) belongs to a class of ferrimagnetism. It has various applications in the form of permanent magnets, microwave absorber, high density magnetic recording media, transformer core, magnetic fluids for biological applications, etc. The crystalline structure of the hexagonal ferrites is the result of close packing of oxygen ion layers. The divalent and trivalent metallic cations are located in interstitial sites of the structure; while the heavy Ba or Sr ions enter substitutionaly the oxygen layers. By replacing the cation and

Materials Research Foundations **142** (2023) 66-92   https://doi.org/10.21741/9781644902318-3

composition with suitable elements, one can prepare different types of ferrites with significant magnetic and electric characteristics. In the literature, it was observed that the effect of ferroelectric cations like Ba, Sr and Pb on the ferrites, hexaferrites and perovskite systems was elucidated in order to study the structural, electrical, optical, magnetic, piezo electric, magnetocaloric and ferroelectric properties. These cations may replace the iron or titanium sites depending upon the preferable site occupation. Moreover, it was observed that the above mentioned properties were changed in a significant way. However, some studies replacing Fe^{3+} ion by other less magnetic moment, paramagnetic or diamagnetic cations can affect the exchange interactions between the magnetic sublattices, which further alter the high uniaxial anisotropy field of the barium hexaferrites. Especially, these Ba and Al cations can influence the magnetic properties of hexaferrites also. However, it is known that the electrical, optical and magnetic properties of ferrites are very sensitive to the particle sizes, shape and degree of crystallinity. At present, tremendous efforts have been made to achieve highly homogeneous nano-particles of barium hexaferrite by using different synthesis methods. Among all synthesis methods, Sol-gel combustion techniques show advantages mainly due to low cost, ability to produce multi-component oxides with single phase and invariably in the nanometer range. Series of 'Cu' doped $BaCuxFe_{12-x}O_{19}$ and 'Al' doped $BaAl_xFe_{12-x}O_{19}$ (x = 0.0, 0.4, 0.8, 1.2) hexaferrites have been prepared separately using sol-gel auto combustion technique. The effects of diamagnetic and paramagnetic elements on structural and magnetic characteristics of $BaFe_{12}O_{19}$ were investigated. The Cu and Al doped M-type barium hexaferrite with chemical composition $BaCuxFe_{12-x}O_{19}$ and $BaAl_xFe_{12-x}O_{19}$ (x = 0.0, 0.4, 0.8, 1.2) were successfully synthesized via sol-gel auto combustion technique. The hexagonal structure of samples remained the same upon doping the copper and aluminium into the barium hexaferrite system. Replacement of Fe^{3+} with different size and nonmagnetic Cu^{2+} and Al^{3+} ions reduced the super exchange interaction and alter the collinear spin alignments. These led to the random variation of saturation and remanence magnetization in substituted samples. Variations in the coercivity values are reversed with the substitution of diamagnetic and paramagnetic elements. The net coercivity decreased with Cu substitution and increased with Al substitution.

Conclusion

This chapter is focused on the historical developments of substituted $BaFe_{12}O_{19}$ hexagonal ferrites and their structure, synthesis methods, various types of properties and applications in different fields. In the introduction, we discussed the various types of properties like structural, electrical and magnetic of substituted $BaFe_{12}O_{19}$ hexagonal ferrites. From the data, a single-phase formation was observed in many substituted $BaFe_{12}O_{19}$ hexagonal ferrites. Magnetic studies revealed the increment resistivity and activation energy with the substitution of different types of dopants and decrease in resistivity with rise in temperature. This increase in resistivity and decrease in dielectric properties make them useful in the microwave devices and the devices operating at higher frequency. Magnetic properties exposed the increase in coercivity make them useful in perpendicular recording media and decrease in coercivity, magnetic moment and magnetic saturation affirm the use of these materials in multilayer chip inductors and beads as high resistivity and low coercivity are basic requirement of these applications. The study of different methods of substituted $BaFe_{12}O_{19}$ hexagonal ferrites confirmed that the results of sol-gel technique and wet chemical method are better than other methods. These kind of substituted $BaFe_{12}O_{19}$ hexagonal ferrites are many applications in computers, radio and television, telecommunication, medical, filters, chemical reactions, etc.

References

[1] Pullar, Robert C, Hexagonal ferrites: A review of the synthesis, properties and applications of hexaferrite ceramics, Progress in Materials Science. 57 (2012) 1191-1334. https://doi.org/10.1016/j.pmatsci.2012.04.001

[2] Chen D, Harward I, Baptist J, Goldman S, Celinski Z, Curie temperature and magnetic properties of aluminum doped barium ferrite particles prepared by ball mill method, Journal of Magnetism and Magnetic Materials. 395 (2015) 350-3. https://doi.org/10.1016/j.jmmm.2015.07.076

[3] Niu X, Liu X, Feng S, Lv F, Huang F, Huang X, Ma Y, Huang K, Effects of presintering temperature on structural and magnetic properties of BaMg1. 8Cu0. 2Fe16O27 hexagonal ferrites, Optik. 126 (2015) 5513-6. https://doi.org/10.1016/j.ijleo.2015.09.044

[4] Rai, G. Murtaza, M. A. Iqbal, and K. T. Kubra, Effect of Ho3+ substitutions on the structural and magnetic properties of BaFe12O19 hexaferrites, Journal of alloys and compounds. 495 (2010) 229-233. https://doi.org/10.1016/j.jallcom.2010.01.133

[5] Mosleh Z, Kameli P, Poorbaferani A, Ranjbar M, Salamati H, Structural, magnetic and microwave absorption properties of Ce-doped barium hexaferrite., Journal of Magnetism and Magnetic Materials. 397 (2016) 101-7. https://doi.org/10.1016/j.jmmm.2015.08.078

[6] Wang L, Zhang J, Zhang Q, Xu N, Song J, XAFS and XPS studies on site occupation of Sm3+ ions in Sm doped M-type BaFe12O19, Journal of Magnetism and Magnetic Materials.377 (2015) 362-7. https://doi.org/10.1016/j.jmmm.2014.10.097

[7] Shams MH, Rozatian AS, Yousefi MH, Valíček J, Šepelák V, Effect of Mg2+ and Ti4+ dopants on the structural, magnetic and high-frequency ferromagnetic properties of barium hexaferrite., Journal of Magnetism and Magnetic Materials.399 (2016) 10-8. https://doi.org/10.1016/j.jmmm.2015.08.099

[8] Shams MH, Rozatian AS, Yousefi MH, Valíček J, Šepelák V, Effect of Mg2+ and Ti4+ dopants on the structural, magnetic and high-frequency ferromagnetic properties of barium hexaferrites, Journal of Magnetism and Magnetic Materials. 399(2016) 10-8. https://doi.org/10.1016/j.jmmm.2015.08.099

[9] Smit, J, H. P. J. Wijn. Ferrites: NV Philips' Gloeilampenfabrieken. Eindhoven, Holland, (1959).

[10] Sözeri H, Deligöz H, Kavas H, Baykal A, Magnetic, dielectric and microwave properties of M-Ti substituted barium hexaferrites (M= Mn2+, Co2+, Cu2+, Ni2+, Zn2+), Ceramics International. 40 (2014) 8645-57. https://doi.org/10.1016/j.ceramint.2014.01.082

[11] Rane MV, Bahadur D, Kulkarni SD, Date SK, Magnetic properties of NiZr substituted barium ferrite, Journal of magnetism and magnetic materials.195 (1999) L256-60. https://doi.org/10.1016/S0304-8853(99)00041-4

[12] Vinnik DA, Zherebtsov DA, Mashkovtseva LS, Nemrava S, Semisalova AS, Galimov DM, Gudkova SA, Chumanov IV, Isaenko LI, Niewa R. Growth, structural and magnetic characterization of Co-and Ni-substituted barium hexaferrite single crystals, Journal of Alloys and Compounds.628 (2015) 480-4. https://doi.org/10.1016/j.jallcom.2014.12.124

[13] Gonzalez-Angeles, A., G. Mendoza-Suárez, A. Gruskova, I. Toth, V. Jančárik, M. Papanova, J. I. Escalante-Garcia, Magnetic studies of NiSn-substituted barium hexaferrites processed by attrition milling, Journal of magnetism and magnetic materials . 270, (2004) 77-83. https://doi.org/10.1016/j.jmmm.2003.08.001

[14] Pullar, R. Hexagonal ferrites, A review of the synthesis, properties and applications of hexaferrite ceramics. Prog. Mater. Sci. 57 (2012) 1191-1334. https://doi.org/10.1016/j.pmatsci.2012.04.001

[15] Lalegani, Zahra, Ali Nemati., Effects of Ce-Co substitution on structural, magnetic and dielectric properties of M-type barium hexaferrite nanoparticles synthetized by sol-gel auto-combustion route. Journal of Materials Science: Materials in Electronics . 26 (2015): 2134-2144. https://doi.org/10.1007/s10854-014-2658-8

[16] Nikmanesh, Hossein, Sedigheh Hoghoghifard, and Behnaz Hadi-Sichani, Study of the structural, magnetic, and microwave absorption properties of the simultaneous substitution of several cations in the barium hexaferrite structure, Journal of Alloys and Compounds. 775 (2019) 1101-1108. https://doi.org/10.1016/j.jallcom.2018.10.051

[17] Sanida, Aikaterini, Sotirios Stavropoulos, Thanassis Speliotis, and Georgios C. Psarras. Magneto-dielectric behaviour of M-Type hexaferrite/polymer nanocomposites. Materials. 11 (2018): 2551. https://doi.org/10.3390/ma11122551

[18] Charoensuk T, Thongsamrit W, Ruttanapun C, Jantaratana P, Sirisathitkul C, Loading effect of sol-gel derived barium hexaferrite on magnetic polymer composites, Nanomaterials. 11(2021) 558. https://doi.org/10.3390/nano11030558

[19] El-Shater, R. E, Exploring the annealing temperature impacting on the magnetic coupling of nanometer soft grain and microstructure hard grain nanocomposites.. Chinese journal of physics .57 (2019) 403-417. https://doi.org/10.1016/j.cjph.2018.10.028

[20] Makled, Mahmoud H, and E. Sheha, An attempt to utilize hard magnetic BaFe12O19 phase as a cathode for magnesium batteries, Journal of Electronic Materials .48 (2019) 1612-1616. https://doi.org/10.1007/s11664-018-06890-0

[21] Borin, Dmitry, Robert Müller, Stefan Odenbach, Magnetoviscosity of a magnetic fluid based on barium hexaferrite nanoplates, Materials. 14 (2021) 1870. https://doi.org/10.3390/ma14081870

[22] Asiri, Synthesis, S. Güner, A. Y. Ş. E. Demir, A. Yildiz, A. Manikandan, and Abdulhadi Baykal, Synthesis and magnetic characterization of Cu substituted barium hexaferrites, Journal of Inorganic and Organometallic Polymers and Materials.28 (2018) 1065-1071. https://doi.org/10.1007/s10904-017-0735-1

[23] Slimani, Y, M. A. Almessiere, A. Baykal, AC susceptibility study of Cu substituted BaFe12O19 nanohexaferrites, Ceramics International. 44(2018): 13097-13105. https://doi.org/10.1016/j.ceramint.2018.04.130

[24] Kumar, Sunil, Shampa Guha, Sweety Supriya, Lagen Kumar Pradhan, and Manoranjan Kar, Correlation between crystal structure parameters with magnetic and dielectric parameters of Cu-doped barium hexaferrites, Journal of Magnetism and Magnetic Materials, 499 (2020) 166213. https://doi.org/10.1016/j.jmmm.2019.166213

[25] Rafiq, M. A, Waqar, M, Muhammad, Q. K, Waleed, M, Saleem, M, Anwar, M. S. Conduction mechanism and magnetic behavior of Cu doped barium hexaferrite

ceramics, Journal of Materials Science: Materials in Electronics. 29(2018) 5134-5142. https://doi.org/10.1007/s10854-017-8477-y

[26] Vadivelan, S, N. Victor Jaya, Investigation of magnetic and structural properties of copper substituted barium ferrite powder particles via co-precipitation method, Results in physics. 6 (2016) 843-850. https://doi.org/10.1016/j.rinp.2016.07.013

[27] Rekaby, M, H. Shehabi, R. Awad. Influence of cobalt addition and calcination temperature on the physical properties of BaFe12O19 hexaferrites nanoparticles, Materials Research Express, 71 (2020) 015057. https://doi.org/10.1088/2053-1591/ab65de

[28] Wang, L., Zhang, J., Zhang, Q., Xu, N., & Song, J. (2015). XAFS and XPS studies on site occupation of Sm^{3+} ions in Sm doped M-type BaFe12O19. Journal of Magnetism and Magnetic Materials, 377, 362-367. https://doi.org/10.1016/j.jmmm.2014.10.097

[29] Bsoul, I, S. H. Mahmood, Magnetic and structural properties of BaFe12− xGaxO19 nanoparticles, Journal of Alloys and Compounds, 489.1 (2010) 110-114. https://doi.org/10.1016/j.jallcom.2009.09.024

[30] Dhage, V. N., Mane, M. L, Babrekar, M. K, Kale, C. M, Jadhav, K. M, Influence of chromium substitution on structural and magnetic properties of BaFe12O19 powder prepared by sol-gel auto combustion method, Journal of Alloys and Compounds. 509(2011) 4394-4398. https://doi.org/10.1016/j.jallcom.2011.01.040

[31] Pawar, R. A., S. S. Desai, Q. Y. Tamboli, Sagar E. Shirsath, S. M. Patange, Ce3+ incorporated structural and magnetic properties of M type barium hexaferrites, Journal of Magnetism and Magnetic Materials, 378 (2015) 59-63. https://doi.org/10.1016/j.jmmm.2014.10.166

[32] Asghar, G, Asri, S, Khusro, S.N, Tariq, G.H, Awan, M.S, Irshad, M., Safeen, A, Iqbal, Y, Shah, W.H, Anis-Ur-Rehman, M, Enhanced Magnetic Properties of Barium Hexaferrite, J. Electron. Mater, 49 (2020) 4318-4323. https://doi.org/10.1007/s11664-020-08125-7

[33] Veisi, S. Shooshtary, M. Yousefi, M. M. Amini, A. R. Shakeri, and M. Bagherzadeh, Magnetic and microwave absorption properties of Cu/Zr doped M-type Ba/Sr hexaferrites prepared via sol-gel auto-combustion method, Journal of Alloys and Compounds .773 (2019) 1187-1194. https://doi.org/10.1016/j.jallcom.2018.09.189

[34] Mudsainiyan, R. K, S. K. Chawla, S. S. Meena, Correlation between site preference and magnetic properties of Co-Zr doped BaCoxZrxFe (12− 2x) O19 prepared under

sol-gel and citrate precursor sol-gel conditions, Journal of alloys and compounds. 615 (2014): 875-881. https://doi.org/10.1016/j.jallcom.2014.07.035

[35] Alzaid, Meshal. Enhancement in optical properties of lanthanum-doped manganese barium hexaferrites under different substitutions." Advances in Condensed Matter Physics. (2021). https://doi.org/10.1155/2021/8849595

[36] Dawar, Naini, Structural, magnetic and dielectric properties of pure and nickel-doped barium nanohexaferrites synthesized using chemical co-precipitation technique, Cogent Physics. 3.1 (2016): 1208450. https://doi.org/10.1080/23311940.2016.1208450

[37] Shao, L.H., Shen, S.Y., Zheng, H., Zheng, P., Wu, Q. Zheng, L, Effect of powder grain size on microstructure and magnetic properties of hexagonal barium ferrite ceramic. Journal of Electronic Materials, 47(2018) 4085-4089. https://doi.org/10.1007/s11664-018-6301-y

[38] Asiri,, Synthesis and magnetic characterization of Cu substituted barium hexaferrites, Journal of Inorganic and Organometallic Polymers and Materials 28.3 (2018) 1065-1071. https://doi.org/10.1007/s10904-017-0735-1

[39] Slimani, Y, M. A. Almessiere, A. Baykal, AC susceptibility study of Cu substituted BaFe12O19 nanohexaferrites. Ceramics International. 44.11 (2018) 13097-13105. https://doi.org/10.1016/j.ceramint.2018.04.130

[40] Kumar, Sunil, Shampa Guha, Sweety Supriya, Lagen Kumar Pradhan, and Manoranjan Kar, Correlation between crystal structure parameters with magnetic and dielectric parameters of Cu-doped barium hexaferrites, Journal of Magnetism and Magnetic Materials. 499 (2020) 166213. https://doi.org/10.1016/j.jmmm.2019.166213

[41] Rafiq, Muhammad Asif, Moaz Waqar, Qaisar Khushi Muhammad, Masam Waleed, Murtaza Saleem, Muhammad Sabieh Anwar, Conduction mechanism and magnetic behavior of Cu doped barium hexaferrite ceramics, Journal of Materials Science: Materials in Electronics . 29(2018) 5134-5142. https://doi.org/10.1007/s10854-017-8477-y

[42] Mohammed, Haetham G., Thar Mohammed Badri Albarody, Susilawati Susilawati, Soheil Gohari, Aris Doyan, Saiful Prayogi, Muhammad Roil Bilad, Reza Alebrahim, and Anwar Ameen Hezam Saeed, Process Optimization of in situ magnetic-anisotropy spark plasma sintering of M-type-based barium hexaferrite BaFe12O19, 14(2021) 2650. https://doi.org/10.3390/ma14102650

[43] Godara, Sachin Kumar, Rahul Kumar Dhaka, Navpreet Kaur, Parambir Singh Malhi, Varinder Kaur, Ashwani Kumar Sood, Shalini Bahel, Synthesis and characterization of

Jamun pulp based M-type barium hexaferrite via sol-gel auto-combustion, Results in Physics. 22 (2021) 103903. https://doi.org/10.1016/j.rinp.2021.103903

[44] Coondoo, Indrani, Neeraj Panwar, Muhammad Asif Rafiq, Venkata S. Puli, Muhammad Nadeem Rafiq, and Ram S. Katiyar, Structural, dielectric and impedance spectroscopy studies in (Bi0. 90R0. 10) Fe0. 95Sc0. 05O3 [R= La, Nd] ceramics, Ceramics International . 7 (2014) 9895-9902. https://doi.org/10.1016/j.ceramint.2014.02.084

[45] Meng, Pingyuan, Kun Xiong, Lin Wang, Shengnan Li, Yankui Cheng, and Guangliang Xu, Tunable complex permeability and enhanced microwave absorption properties of BaNixCo1− xTiFe10O19, Journal of Alloys and Compounds. 628 (2015) 75-80. https://doi.org/10.1016/j.jallcom.2014.10.163

[46] Rafiq, Muhammad Asif, Alexander Tkach, Maria Elisabete Costa, Paula Maria Vilarinho., Defects and charge transport in Mn-doped K 0.5 Na 0.5 NbO 3 ceramics, Physical Chemistry Chemical Physics , 37 (2015) 24403-24411. https://doi.org/10.1039/C5CP02883C

[47] Sözeri, Hüseyin, Effect of pelletization on magnetic properties of BaFe12O19, Journal of alloys and compounds, 486 (2009): 809-814. https://doi.org/10.1016/j.jallcom.2009.07.072

[48] Topal, Ugur, Halil I. Bakan. Magnetic properties and remanence analysis in permanently magnetic BaFe12O19 foams, Journal of the European Ceramic Society, 30.15 (2010) 3167-3171. https://doi.org/10.1016/j.jeurceramsoc.2010.06.008

[49] Garcia-Casillas, P. E., A. M. Beesley, D. Bueno, J. A. Matutes-Aquino, and C. A. Martinez, Remanence properties of barium hexaferrites, Journal of alloys and compounds, 369 (2004) 185-189. https://doi.org/10.1016/j.jallcom.2003.09.100

[50] Dho, Joonghoe, E. K. Lee, J. Y. Park, N. H. Hur, Effects of the grain boundary on the coercivity of barium ferrite BaFe12O19, Journal of Magnetism and Magnetic Materials. 285 (2005) 164-168. https://doi.org/10.1016/j.jmmm.2004.07.033

[51] Rashad, M. M., M. Radwan, M. M. Hessien, Effect of Fe/Ba mole ratios and surface-active agents on the formation and magnetic properties of co-precipitated barium hexaferrites, Journal of Alloys and Compounds. 453 (2008) 304-308. https://doi.org/10.1016/j.jallcom.2006.11.080

[52] Topal, Ugur, Husnu Ozkan, Huseyin Sozeri, Synthesis and characterization of nanocrystalline BaFe12O19 obtained at 850 C by using ammonium nitrate melt,

Journal of magnetism and magnetic materials. 284 (2004) 416-422.
https://doi.org/10.1016/j.jmmm.2004.07.009

[53] Kerschla, P. R. Gr. ossingerb, C. Kussbachb, R. Sato-Turtellib, KHM Ullera, L. Schultza. J Magn Magn Mater. (2002) 1468-1470.

[54] Martirosyan, K. S., E. Galstyan, S. M. Hossain, Yi-Ju Wang, D. Litvinov, Barium hexaferrite nanoparticles: synthesis and magnetic properties, Materials Science and Engineering, 176 (2011): 8-13. https://doi.org/10.1016/j.mseb.2010.08.005

[55] Handoko, Erfan. Effect of Co-Ti Substitution on Magnetic Properties of Nanocrystalline BaFe12O19." KnE Engineering , 2016.
https://doi.org/10.18502/keg.v1i1.506

[56] Dhage, Vinod N., M. L. Mane, A. P. Keche, C. T. Birajdar, K. M. Jadhav, Structural and magnetic behaviour of aluminium doped barium hexaferrite nanoparticles synthesized by solution combustion technique, Physica B: Condensed Matter ,.406 (2011) 789-793. https://doi.org/10.1016/j.physb.2010.11.094

[57] Mariño-Castellanos, P. A, J. C. Somarriba-Jarque, J. Anglada-Rivera, Magnetic and microstructural properties of the BaFe (12−(4/3) x) SnxO19 ceramic system, Physica B: Condensed Matter.362 (2005) 95-102. https://doi.org/10.1016/j.physb.2005.01.480

[58] Gonzalez-Angeles, A, G. Mendoza-Suárez, A. Gruskova, I. Toth, V. Jančárik, M. Papanova, J. I. Escalante-Garcıa, Magnetic studies of NiSn-substituted barium hexaferrites processed by attrition milling, Journal of magnetism and magnetic materials . 270 (2004) 77-83. https://doi.org/10.1016/j.jmmm.2003.08.001

[59] Mariño-Castellanos, P. A., J. Anglada-Rivera, A. Cruz-Fuentes, R. Lora-Serrano "Magnetic and microstructural properties of the Ti4+-doped Barium hexaferrites, Journal of Magnetism and Magnetic Materials .280 (2004) 214-220.
https://doi.org/10.1016/j.jmmm.2004.03.015

[60] Gruskova, A, J. Slama, R. Dosoudil, D. Kevicka, V. Jančárik, I. Toth, Influence of Co-Ti substitution on coercivity in Ba ferrites, Journal of magnetism and Magnetic Materials. 242 (2002) 423-425. https://doi.org/10.1016/S0304-8853(01)01139-8

[61] Ghasemi, A., A. Hossienpour, A. Morisako, A. Saatchi, M. Salehi. Electromagnetic properties and microwave absorbing characteristics of doped barium hexaferrites, Journal of Magnetism and Magnetic Materials. 302 (2006)429-435.
https://doi.org/10.1016/j.jmmm.2005.10.006

[62] Qiu, Jianxun, Haigen Shen, Mingyuan Gu, Microwave absorption of nanosized barium ferrite particles prepared using high-energy ball milling, Powder Technology. 154(2005) 116-119. https://doi.org/10.1016/j.powtec.2005.05.003

[63] Tenaud, P., A. Morel, F. Kools, J. M. Le Breton, L. Lechevallier. Recent improvement of hard ferrite permanent magnets based on La-Co substitution, Journal of alloys and compounds . 370 (2004) 331-334. https://doi.org/10.1016/j.jallcom.2003.09.106

[64] Corral-Huacuz, J. C, G, Mendoza-Suarez, Preparation and magnetic properties of Ir-Co and La-Zn substituted barium ferrite powders obtained by sol-gel, Journal of magnetism and magnetic materials. 242 (2002) 430-433. https://doi.org/10.1016/S0304-8853(01)01141-6

[65] Polyko, D. D., L. A. Bashkirov, S. V. Trukhanov, L. S. Lobanovskii, I. M. Sirota, Crystal structure and magnetic properties of high-coercivity Sr1− x Pr x Fe12− x Zn x O19 solid solutions, Inorganic Materials .47 (2011): 75-79. https://doi.org/10.1134/S0020168511010110

[66] Alsmadi, A. M., I. Bsoul, S. H. Mahmood, G. Alnawashi, F. M. Al-Dweri, Y. Maswadeh, U. Welp, Magnetic study of M-type Ru-Ti doped strontium hexaferrite nanocrystalline particles, Journal of Alloys and Compounds. 648 (2015) 419-427. https://doi.org/10.1016/j.jallcom.2015.06.274

[67] Pullar, Robert C, Hexagonal ferrites: a review of the synthesis, properties and applications of hexaferrite ceramics, Progress in Materials Science. 57.7 (2012) 1191-1334. https://doi.org/10.1016/j.pmatsci.2012.04.001

[68] Wagner, T. R, Preparation and crystal structure analysis of magnetoplumbite-type BaGa12O19, Journal of Solid State Chemistry. 136.1 (1998): 120-124. https://doi.org/10.1006/jssc.1997.7681

[69] Chawla, S. K., S. S. Meena, Prabhjoyt Kaur, R. K. Mudsainiyan, S. M. Yusuf, Effect of site preferences on structural and magnetic switching properties of CO-Zr doped strontium hexaferrite SrCoxZrxFe (12− 2x) O19, Journal of Magnetism and Magnetic Materials.378 (2015) 84-91. https://doi.org/10.1016/j.jmmm.2014.10.168

[70] Kumar, Rajnish, and Manoranjan Kar, Correlation between lattice strain and magnetic behavior in non-magnetic Ca substituted nano-crystalline cobalt ferrite, Ceramics International. 42.6 (2016) 6640-6647. https://doi.org/10.1016/j.ceramint.2016.01.007

[71] Ahmad, Syed Ismail, Shakeel Ahmed Ansari, D. Ravi Kumar, Structural, morphological, magnetic properties and cation distribution of Ce and Sm co-substituted nano crystalline cobalt ferrite, Materials Chemistry and Physics. 208 (2018): 248-257. https://doi.org/10.1016/j.matchemphys.2018.01.050

[72] Tauc, J, Radu Grigorovici, Anina Vancu, Optical properties and electronic structure of amorphous germanium, physica status solidi (b). 15.2 (1966) 627-637. https://doi.org/10.1002/pssb.19660150224

[73] Baykal, Abdulhadi, S. Guner, H. Gungunes, K. M. Batoo, Md Amir, A. Manikandan, Magneto optical properties and hyperfine interactions of Cr^{3+} ion substituted copper ferrite nanoparticles., Journal of Inorganic and Organometallic Polymers and Materials .28 (2018) 2533-2544. https://doi.org/10.1007/s10904-018-0903-y

[74] Baykal, A. Y. Ş. E., S. Esir, A. Demir, S. Güner, Magnetic and optical properties of $Cu1-xZnxFe2O4$ nanoparticles dispersed in a silica matrix by a sol-gel auto-combustion method, Ceramics International. 41 (2015): 231-239. https://doi.org/10.1016/j.ceramint.2014.08.063

[75] Manikandan, A., R. Sridhar, S. Arul Antony, Seeram Ramakrishna, A simple aloe vera plant-extracted microwave and conventional combustion synthesis: morphological, optical, magnetic and catalytic properties of $CoFe2O4$ nanostructures, Journal of Molecular Structure. 1076 (2014): 188-200. https://doi.org/10.1016/j.molstruc.2014.07.054

[76] Manikandan, A., M. Durka, K. Seevakan, S. Arul Antony, A novel one-pot combustion synthesis and opto-magnetic properties of magnetically separable spinel $Mn x Mg 1- x Fe2 O 4$ ($0.0\leq x\leq 0.5$) nanophotocatalysts., Journal of Superconductivity and Novel Magnetism .28 (2015): 1405-1416.. https://doi.org/10.1007/s10948-014-2864-x

[77] Ahmed, Arham S, Band gap narrowing and fluorescence properties of nickel doped $SnO2$ nanoparticles, Journal of luminescence, 131.1 (2011): 1-6. https://doi.org/10.1016/j.jlumin.2010.07.017

[78] Stoner, Edmund Clifton, E. P. Wohlfarth, A mechanism of magnetic hysteresis in heterogeneous alloys, Philosophical Transactions of the Royal Society of London. Series A, Mathematical and Physical Sciences.826 (1948) 599-642. https://doi.org/10.1098/rsta.1948.0007

[79] Slimani, Yassine, Abdulhadi Baykal, Md Amir, N. Tashkandi, Hakan Güngüneş, S. Guner, H. S. El Sayed, F. Aldakheel, Tawfik A. Saleh, A. Manikandan, Substitution

effect of Cr3+ on hyperfine interactions, magnetic and optical properties of Sr-hexaferrites, Ceramics International, 44 (2018) 15995-16004.
https://doi.org/10.1016/j.ceramint.2018.06.033

[80] Auwal, I. A., Hakan Güngüneş, Abdulhadi Baykal, Sadık Güner, Sagar E. Shirsath, and Murat Sertkol, Structural, morphological, optical, cation distribution and Mössbauer analysis of Bi3+ substituted strontium hexaferrites, Ceramics International. 42(2016): 8627-8635. https://doi.org/10.1016/j.ceramint.2016.02.094

[81] Auwal, I. A., A. Baykal, S. Güner, and H. Sözeri, Magneto-optical properties of SrBixLaxFe12-2xO19 (0.0≤ x≤ 0.5) hexaferrites by sol-gel auto-combustion technique, Ceramics Internationa. 43(2017): 1298-1303.
https://doi.org/10.1016/j.ceramint.2016.10.080

[82] Dom, Rekha, Pramod H. Borse, C. R. Cho, J. S. Lee, S. M. Yu, J. H. Yoon, T. E. Hong, E. D. Jeong, and H. G. Kim, Synthesis of SrFe 12 O 19 and Sr 7 Fe 10 O 22 systems for visible light photocatalytic studies, Journal of Ceramic Processing Research, 13(2012) 451-456.

[83] Javidan, Abdollah, Somayeh Rafizadeh, S. Mostafa Hosseinpour-Mashkani, Strontium ferrite nanoparticle study: thermal decomposition synthesis, characterization, and optical and magnetic properties, Materials science in semiconductor processing. 27 (2014) 468-473.
https://doi.org/10.1016/j.mssp.2014.07.024

[84] Güner, Synthesis, I. A. Auwal, A. Baykal, and H. Sözeri, Synthesis, characterization and magneto optical properties of BaBixLaxYxFe12- 3xO19 (0.0≤ x≤ 0.33) hexaferrites, Journal of Magnetism and Magnetic Materials, 416 (2016): 261-268.
https://doi.org/10.1016/j.jmmm.2016.04.091

[85] Almessiere, M. A., Y. Slimani, M. Sertkol, M. Nawaz, A. Baykal, I. Ercan, The impact of Zr substituted Sr hexaferrite: investigation on structure, optic and magnetic properties. Results in Physics .13 (2019) 102244.
https://doi.org/10.1016/j.rinp.2019.102244

[86] Javidan, Abdollah, Somayeh Rafizadeh, S. Mostafa Hosseinpour-Mashkani., Strontium ferrite nanoparticle study: thermal decomposition synthesis, characterization, and optical and magnetic properties, Materials science in semiconductor processing.27 (2014) 468-473.
https://doi.org/10.1016/j.mssp.2014.07.024

[87] Güner, Synthesis, I. A. Auwal, A. Baykal, H. Sözeri, Synthesis, characterization and magneto optical properties of BaBixLaxYxFe12- 3xO19 (0.0≤ x≤ 0.33) hexaferrites."

Journal of Magnetism and Magnetic Materials . 416 (2016) 261-268.2
https://doi.org/10.1016/j.jmmm.2016.04.091

[88] Almessiere, M. A., Y. Slimani, M. Sertkol, M. Nawaz, A. Baykal, and I. Ercan..The
impact of Zr substituted Sr hexaferrite: investigation on structure, optic and magnetic
properties, Results in Physics, 13 (2019) 102244..
https://doi.org/10.1016/j.rinp.2019.102244

[89] Almessiere, Munirah Abdullah, Yassine Slimani, H. Gungunes, A. Manikandan, and
Abdulhadi Baykal, Investigation of the effects of Tm3+ on the structural,
microstructural, optical, and magnetic properties of Sr hexaferrites,Results in Physics.
13 (2019) 102166. https://doi.org/10.1016/j.rinp.2019.102166

[90] Amir, Md, A. Baykal, S. Güner, M. Sertkol, H. Sözeri, and Muhammet Toprak.,
Synthesis and characterization of CoxZn1− xAlFeO4 nanoparticles, Journal of
Inorganic and Organometallic Polymers and Materials.25 (2015) 747-754.
https://doi.org/10.1007/s10904-014-0153-6

[91] Auwal, I. A., Hakan Güngüneş, Sadık Güner, Sagar E. Shirsath, Murat Sertkol, and
Abdulhadi Baykal, Structural, magneto-optical properties and cation distribution of
SrBixLaxYxFe12− 3xO19 (0.0≤ x≤ 0.33) hexaferrites, Materials Research Bulletin .80
(2016) 263-272. https://doi.org/10.1016/j.materresbull.2016.03.028

[92] Almessiere, M. A., Y. Slimani, H. S. El Sayed, A. Baykal, and I. Ercan,
Microstructural and magnetic investigation of vanadium-substituted Sr-
nanohexaferrite, Journal of Magnetism and Magnetic Materials. 471 (2019) 124-132.
https://doi.org/10.1016/j.jmmm.2018.09.054

[93] Almessiere, M. A., Y. Slimani, and A. Baykal, Structural and magnetic properties of
Ce-doped strontium hexaferrites, Ceramics International, 44.8 (2018) 9000-9008.
https://doi.org/10.1016/j.ceramint.2018.02.101

[94] Mubarak, Tahseen II., Olfat A. Mahmood, and Zahraa J. Hamakhan, Structural,
magnetic and electrical properties, International Journal of Applied Engineering
Research. 13.8 (2018) 6369-6379.

[95] Ashiq, Muhammad Naeem, Raheela Beenish Qureshi, Muhammad Aslam Malana,
and Muhammad Fahad Ehsan., Synthesis, structural, magnetic and dielectric properties
of zirconium copper doped M-type calcium strontium hexaferrites, Journal of alloys
and compounds. 617 (2014) 437-443.. https://doi.org/10.1016/j.jallcom.2014.08.015

[96] Wang, Yuping, Liangchao Li, Hui Liu, Haizhen Qiu, and Feng Xu, Magnetic properties and microstructure of La-substituted BaCr-ferrite powders, Materials Letters. 62 (2008) 2060-2062. https://doi.org/10.1016/j.matlet.2007.11.026

[97] Kaur, Prabhjyot, S. K. Chawla, Sukhleen Bindra Narang, and Kunal Pubby, Effect of Cu-Co-Zr doping on the properties of strontium hexaferrites synthesized by sol-gel auto-combustion method, Journal of Superconductivity and Novel Magnetism .30(2017): 635-645. https://doi.org/10.1007/s10948-016-3835-1

[98] Raghuvanshi, S., P. Tiwari, S. N. Kane, D. K. Avasthi, F. Mazaleyrat, Tetiana Tatarchuk, and Ivan Mironyuk., Dual control on structure and magnetic properties of Mg ferrite: role of swift heavy ion irradiation, Journal of Magnetism and Magnetic Materials .471 (2019) 521-528. https://doi.org/10.1016/j.jmmm.2018.10.004

[99] El-Fadl, A. Abu, A. M. Hassan, M. H. Mahmoud, Tetiana Tatarchuk, I. P. Yaremiy, A. M. Gismelssed, and M. A. Ahmed, Synthesis and magnetic properties of spinel Zn1− xNixFe2O4 (0.0≤ x≤ 1.0) nanoparticles synthesized by microwave combustion method., Journal of Magnetism and Magnetic Materials. 471 (2019): 192-199. https://doi.org/10.1016/j.jmmm.2018.09.074

[100] Tiwari, P., R. Verma, S. N. Kane, Tetiana Tatarchuk, and F. Mazaleyrat., Effect of Zn addition on structural, magnetic properties and anti-structural modeling of magnesium-nickel nano ferrites, Materials Chemistry and Physics .229 (2019) 78-86. https://doi.org/10.1016/j.matchemphys.2019.02.030

[101] Alsmadi, A. M., I. Bsoul, S. H. Mahmood, G. Alnawashi, K. Prokeš, K. Siemensmeyer, B. Klemke, and H. Nakotte, Magnetic study of M-type doped barium hexaferrite nanocrystalline particles, Journal of Applied Physics. 24 (2013) 243910. https://doi.org/10.1063/1.4858383

[102] Zi, Z. F., X. H. Ma, Y. Y. Wei, Q. C. Liu, M. Zhang, X. B. Zhu, and Y. P. Sun, Influence of La-Mn substitutions on magnetic properties of M-type strontium hexaferrites. AIP Advances. 8(2018) 056235. https://doi.org/10.1063/1.5007695

[103] J. Liu, Y. Zeng, X. Zhang, M. Zhang, Effects of magnetic pre-alignment of nanopowders on formation of high textured barium hexa-ferrite quasi-single crystals via a magnetic forming and liquid participation sintering route, J. Magn. Magn. Mater. 382 (2015) 188-192. https://doi.org/10.1016/j.jmmm.2015.01.078

[104] H.M. Khan, M.U. Islam, Y. Xiu, M.A. Iqbal, I. Ali, Structural and magnetic properties of TbZn-substituted calcium barium M-type nano-structured hexa-ferrites, J. Alloy. Compd. 589 (2014) 258-262. https://doi.org/10.1016/j.jallcom.2013.11.107

[105] Y. Slimani, A. Baykal, A. Manikandan, Effect of Cr3+ substitution on AC susceptibility of Ba hexaferrite nanoparticles, J. Magn. Magn. Mater. 458 (2018) 204-212. https://doi.org/10.1016/j.jmmm.2018.03.025

[106] V.G. Harris, A. Geiler, Y. Chen, S.D. Yoon, M. Wu, A. Yang, Z. Chen, P. He, P.V. Parimi, X. Zuo, Recent advances in processing and applications of microwave ferrites, J. Magn. Magn. Mater. 321 (2009) 2035-2047. https://doi.org/10.1016/j.jmmm.2009.01.004

[107] J.F. Wang, C.B. Ponton, R. Grossinger, I.R. Harris, A study of La-substituted strontium hexaferrite by hydrothermal synthesis, J. Alloy. Compd. 369 (2004) 170. https://doi.org/10.1016/j.jallcom.2003.09.097

[108] L. Lechevallier, J.M. LeBreton, J.F. Wang, I.R. Harris, Structural analysis of hydrothermally synthesized $Sr1-xSmxFe12O19$ hexagonal ferrites, J. Magn. Magn. Mater. 269 (2004) 192. https://doi.org/10.1016/S0304-8853(03)00591-2

[109] N. Rezlescu, C. Doroftei, E. Rezlescu, P.D. Popa, Fine-grained erbium-doped strontium hexaferrite, Phys. Status Solidi (a). 203 (2006) 3844-3851. https://doi.org/10.1002/pssa.200622213

[110] C. Doroftei, E. Rezlescu, P.D. Popa, N. Rezlescu, Heat-treatment influence on the microstructure and magnetic properties of rare-earth substituted $SrFe12O19$, J. Cryst. Res. Technol. 41 (2006) 1112-1119. https://doi.org/10.1002/crat.200610731

[111] M.A. Ahmed, E. Ateia, S.I. El-Dek, F.M. Salem, Rate of heating and sintering temperature effect on the electrical properties of Nd ferrite, J. Mater. Sci. 38 (2003) 1087. https://doi.org/10.1023/A:1022314301113

[112] Packiaraj, G., Hashim, M., Naidu, K.C.B., Joice, G.H.R., Naik, J.L., Ravinder, D. and Rao, B.R., 2020. Magnetic properties of Cu and Al doped nano $BaFe12O19$ ceramics. Biointerface Res. Appl. Chem, 10, pp.5455-5459. https://doi.org/10.33263/BRIAC103.455459

An Introduction to Hard Ferrites: From Fundamentals to Practical Applications   Materials Research Forum LLC
Materials Research Foundations **142** (2023) 93-120   https://doi.org/10.21741/9781644902318-4

Chapter 4

Effect of Substitution on the Electric and Magnetic Properties of $SrFe_{12}O_{19}$ Hexa Hard Ferrites

Dipti Rawat, Ragini Raj Singh*

Nanotechnology Laboratory, Department of Physics and Materials Science, Jaypee University of Information technology, Waknaghat Solan 173234, H.P., India

* raginirajsingh@gmail.com

Abstract

The hexagonal ferrites, also known as hexaferrites, have stimulated interest subsequent to their finding in the 1950s, which is steadily growing today. Both commercially and technologically, these materials have grown in importance. In addition to their employment as permanent magnets, they are commonly used as "magnetic recording and data storage materials, as well as components in electrical systems, notably those that activates at microwave/GHz frequency range". The goal of the presented study is to offer new ideas for the development of magnetic samples Strontium hexaferrite $SrFe_{12}O_{19}$ (SrM) that are suited for specific applications, as well as to explain the influence of rare-earth (RE) substitution on the magnetic and electrical properties of SrM.

Keywords

Hexaferrite, Rare-Earth, Electrical Properties, Magnetic Properties, Substitutions

Contents

1.   Introduction

Due to their unique size-dependent features and wide variety of applications in science and technology, nanoscale SrM are set up as promising material [1-7]. Due to expanding uses in the construction of electrical components for automobiles and wireless communication devices, scientific attention in SrM magnetic nanoparticles (MNPs) has lately been reignited. Furthermore, SrM is employed in bonded magnets, different microwave devices (filters, phase shifters, isolators, circulators), and "high-density magnetic" and "magneto-optic recorders" [8–10]. Crystal structure for the SrM designed using vista software is inscribed in Fig. 1.

SrM's hexagonal structure is considered to be fabricated of alternating spinel ($S = Fe_6O_8^{2+}$) and hexagonal ($R = SrFe_6O_{11}^{2-}$) layers. O^{2-} ions and Sr^{2+} ions are closed packed, in the hexagonal layer and, the Fe^{3+} ions are spread at the sites "octahedral" (12k, 2a and 4f2), "trigonal-bipyramidal" (2b) and "tetrahedral" (4f1). Super-exchange interactions through the O^{2-} ions assist magnetic moments of the Fe^{3+} ions coupled to each other. The Sr^{2+} ion is much accountable for the big "magnetic uniaxial anisotropy" as it causes agitation of the crystal lattice [11]. Computer memory chips, high-density recording media, transformers, microwave devices, and permanent magnets are examples of industrial applications of SrM [12-14]. It has been noted that hexagonal ferrites' intrinsic features like there saturation magnetization, high coercivity there uniaxial anisotropy all plays a significant role in the fabrication of electronic, magnetic and microwave device components. Due to its high coercivity, which stems from its high magnetocrystalline anisotropy and is highly reliant on particle size and shape, $SrFe_{12}O_{19}$ is a hard magnetic material. SrM has been widely employed in industrial applications as permanent magnets because it is easily powdered and shaped into required shapes. Strontium has a lower toxicity than lead and barium. As a result, SrM ferrites are being used in non-toxic micro and nano systems such as biomarkers, biodiagnostics, and biosensors [15].

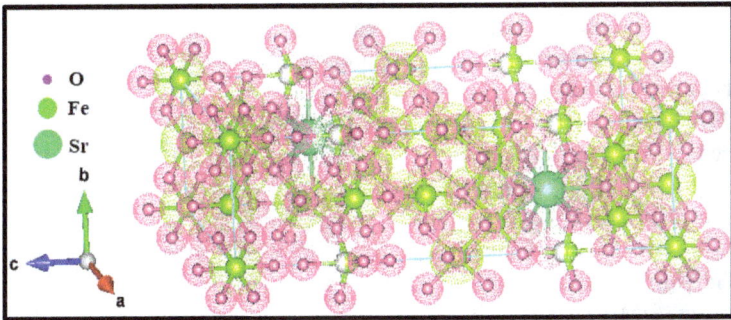

Figure 1. Crystal structure for the SrM.

Chemical composition, synthesis technique, calcination temperature, and even particle size has all been shown to affect the electrical, dielectric, and magnetic properties of ferrites [16-19]. As a result, investigating these variables may yield trustworthy information regarding hexagonal ferrites' intrinsic features. Not just from a fundamental standpoint, but also from a practical standpoint, dielectric and electrical properties are just as significant as magnetic qualities for hexaferrites. Rare earth element substitution has been found to improve the electro-magnetic characteristics of hexaferrites in recent investigations [13-14]. The increased features can be attributed to RE ions' startling properties, such as high resistivity and unpaired 4f electrons that cause significant spin–orbit angular momentum coupling. The magnetic characteristics of SrM have been reported to change when RE is substituted singularly or in combination with transition metals such as "La, Pr, Nd, Ni-Er, La-Zn, La-Co, Nd-Co, Pr-Co" [12]. RE ions have a low solubility in Sr-hexaferrites, and their introduction causes secondary phases to develop, which essentially be evaded for the purpose of attaining permanent magnets with ideal characteristics. Due to their required magnetic characteristics, ferrite magnets may not be as powerful as RE magnets (SmCo and NdFeB) and RE alloys [20]. In comparison to commercial NdFeB (583 K), SrM has a higher Curie temperature of 733 K [12]. Due to their low production costs, they are also the most extensively utilised magnets. High remanence, high coercivity, and a big energy product $(BH)_{max}$,SrM is a good choice for use in magnetic water treatment. Small particle size, growth anisotropy, and high-density ferrite are required to achieve these qualities. The improvement of coercivity is more difficult than improving the energy product $(BH)_{max}$. The magnetic characteristics can be improved by using a greater density ferrite with a uniform grain distribution. Researchers are now seeking to investigate the magnetic properties of the material by altering the stoichiometry, chemical purity, and production conditions. The best technique to identify productive compositions for varied purposes is to substitute the Sr^{2+} and Fe^{3+} cations. Different cations, such as Ho^{3+}, Ti^{4+}, Al^{3+}, Cr^{3+}, and Ga^{3+}, have been used to replace Fe^{3+}ions in strontium hexaferrite in previous studies [21-22]. Other elements, such as La^{3+}, Nd^{3+}, Sm^{3+}, Pr^{3+}, and Gd^{3+}, are used to replace some of them [23-25]. Following that, different synthesis methods were used to successfully achieve coupled substitution (such as Mn–Sn–Ti, Zn–Nb, and La–Cu) in M-type hexagonal ferrites. The sol–gel process, co-precipitation method, self-propagation, mechanical alloying methods, microwave, hydrothermal, and ultrasound-assisted synthesis are the main ways for making strontium hexaferrite. The sol-gel method was used to prepare $Sr(Ho_xGd_ySm_z)Fe_{(12-(x+z+y))}O_{19}$ (x=y=z=0.01) by doping tiny amounts of Sm^{3+}, Gd^{3+}, and Ho^{3+} ions into SrM concurrently [26]. The magnetic characteristics of $Sr(Ho_xGd_ySm_z)Fe_{(12-(x+z+y))}O_{19}$ (x=y=z=0.01) (labelled RE-SrM) were examined as well as the Sm^{3+}, Gd^{3+}, and Ho^{3+} ion replacement mechanisms [26]. The influence of substitution on hard ferrite attributes such as electric and magnetic properties is the focus of this chapter. The magnetic characteristics of this RE doped SrM suggested that it could be employed in permanent magnets as well as water treatment. Further the hard/soft composites $SrTb_{0.01}Tm_{0.01}Fe_{11.98}O_{19}/(CoFe_2O_4)$ and $0.7BaTiO_3–0.3SrFe_{12}O_{19}$ for the magnetic and electrical properties are also discussed.

2.   Synthesis technique for hexagonal hard ferrites

Despite the fact that numerous researchers have studied the creation of hexagonal ferrites for over 50 years [27-30], the mechanisms involved are still unknown. To become the main phase, they all require high temperatures of at least 1000 °C. Due to the high temperature requirement, pure ferrites with only one phase are extremely difficult to acquire. Sometimes in ferrites having only one pure phase, at temperatures above 1200 °C, an exaggerated growth outline is observed, in which certain particles develop to a considerably bigger extent and at a much faster rate than their nearby particles. Single hexagonal grains with a diameter of up to 1 mm can form in this morphology, which contains a wide range of particle diameters ranging from tens to hundreds of μm. This morphology is known as discontinuous grain growth (DGG). Section 2.1 onward delves more into the solid-state processes and techniques of production of certain ferrites.

2.1 Standard ceramic techniques

The hexagonal ferrites were made using standard ceramic procedures, which involved heating a mixture of oxide and Ba or Sr carbonate salt to obtain the required phase. To make a finer material, the ceramic product is frequently milled and powdered, and then annealed at higher temperature to increase density. Because of the high temperatures and extended firing times required, coarse grain products are typically produced, with average grain sizes ranging from 1 to 10 μm for M ferrites and up to 50 μm for higher hexagonal phases like Y, Z, and W. In most of the ferrites like $BaFe_{12}O_{19}$ at 1000 °C single phase formation is attained [31]. Crystallisation happens at a lower temperature, leading in smaller particles and lower sintering temperatures, and the fully densified material can typically be formed in one step if the beginning materials are intimately combined at the atomic or ionic level before reaction. This can be done with high-energy mechanochemical milling, which is generally done in a high-energy planetary ball mill. Because of the small particle size and milling energy, this not only produces finely dispersed powders that can be nanosized after extensive milling, but it also allows for chemical assembly with significantly lower activation energies. Hexaferrites are good for this form of synthesis since they are made up of pre-existing crystalline building blocks. This is especially true for more complicated molecules. Converting ferrites to an ultrafine dispersed solid-state of nano agglomerates or nanoclusters enables nanoparticles (NPs) to engage in thermal motion and allows for their self-assembly into superstructures by eventually achieving the thermodynamic optimum [32]. Several more approaches to synthesise hexaferrites, and the bulk of those are considered in the sections below which rely on the precipitation of a ferrite MNPs precursor at a particular point to achieve hexaferrite synthesis with varied degrees of success.

2.2 Co-precipitation

Since the early 1960s, chemical co-precipitation of salts with a base has been employed to generate ferrites, resulting in a precipitate having all of the components combined at an ionic level. It has been discovered that an iron deficient non-stoichiometric mixture must be used instead of the right ratio of 12 [33], for example with a Fe:Ba ratio of 10–11 for Barium ferrite (BaM). To produce pure M from co-precipitates, the Fe:Ba ratio must be smaller than 9. BaM can be formed with a submicron grain size at temperatures between 750 °C and 900 °C using this process; centrifugation can enhance the density of sample rather than as that of decantation [34] if washed several times. Wet chemical technique was advanced to increase uniformity, which uses an aqueous solution of metal salts to form precipitate with a strong base, resulting in hydroxide which was oxidised by sparkling air through the suspension to produce a homogeneous fine-grained MNPs. Recently, non-stoichiometric solutions of iron (II) salts and barium salts were co-precipitated with NaOH at pH 14 to produce hydroxides, which were then oxidised with hydrogen peroxide (H_2O_2) and washed to pH 7. The dried hydroxides are heated to produce pure BaM with a grain size of less than 0.1 μm [35] at a very low temperature of 600 °C. BaM has also been produced using sodium hypochlorite (NaClO) for a similar purpose. Co_2Z was created by co-precipitating very dilute solutions of barium and cobalt chlorides, as well as iron nitrate, then reacting the precipitates with acetic acid and an organic stabiliser, filtering them, and finally processing them by traditional ceramic techniques to get Co_2Z [12].

2.3   Sol-Gel

The sol–gel technique combines metal-organic precursor or inorganic particles on a colloidal scale resulting in fine-grained polycrystalline ferrites with a limited size distribution. An aqueous solution of metal salts is co-precipitated by a base, but instead of drying and burning the precipitates, they are treated to form a colloidal sol, which can then be condensed to a gel and then burned to produce ferrites [36]. To make a sol, an organic coordinating agent like ethylene glycol is commonly added to the hydroxide solution, which forms a gel structure after the water evaporates. When BaM was created by evaporating glycol containing co-precipitated salts to produce a homogeneous gel, it was discovered that a non-stoichiometric combination was required, and a Fe:Ba ratio of 10.5 produced M ferrite at 900 °C /1 h with a grain size of just 200 nm [37]. Ample Ba containing precursor only provided pure M with no α-Fe_2O_3 as a secondary phase, but with a ratio of less than nine $BaFe_2O_4$ occurred as secondary phase in its place, showing that the production of BaM by this approach has a limited compositional window. At 60°C, washed iron (III) hydroxide was dissolved in citric acid solution, after which $BaCO_3$ in the molar ratio of Fe:Ba = 11.6 was added, and then viscous residue was heated until it changes to gel at 170°C. After treating to eliminate the organic components, single phase BaM with hexagonal crystals in a size range of 90–110 nm was formed at 750°C, with additional annealing having minimal influence on the grain size [38]. The stoichiometric SHF precursor synthesised through a NO_3-based sol was extremely even, with 8 nm of average sol particle size and a concentration of up to 17% Fe^{3+}, whereas the BaM precursor sol made from nitrates was poly disperse and much less stable, with an average sol particle size of 54 nm and an upper limit of 282 nm [39]. All of the hexagonal ferrites named above were made into very stable stoichiometric sol precursors via nitrate-free sol–gel environment with the addition of cobalt halide salt, with 10 nm of average sol particle sizes, and all of the MNPs were transformed into the single phase hexaferrites upon heating amid 700°C (SrM) and 1200°C (Co$_2$W and Co$_2$Z). Due to the near absolute insolubility of barium sulphate, it was discovered that no barium-containing sols could be made using any sulphate salts or a SO$_4^{-2}$ or SO$_3^{-2}$ stabilised iron sol. Some other techniques for the synthesis of ferrites are shown in Fig. 2.

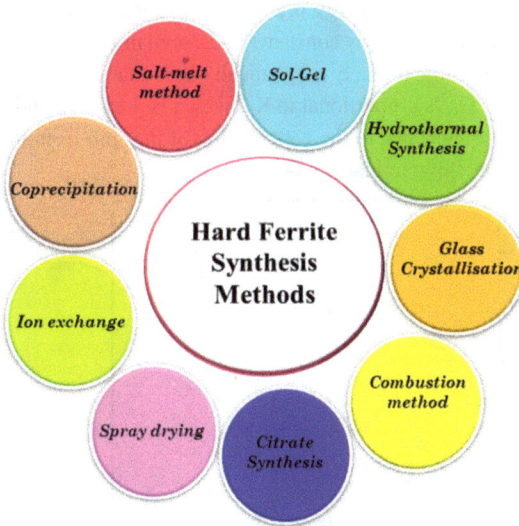

Figure 2. Synthesis methods for hard magnetic nanoparticles.

3.　Magnetism in hexagonal ferrites

The "magneto crystalline anisotropy (MCA)" in hexaferrites is caused by the presence of at least one big metal 2+ ion (typically Sr^{2+} or Ba^{2+}), which creates a minor disruption in the lattice due to size changes. Because the c-axis is the preferred magnetisation axis for hexagonal ferrites, loose crystals align themselves with the c-axis parallel to the field, resulting in distinct XRD pattern of hexagonal ferrite than randomly oriented examples. When the magnetic characteristics are tested in the direction of alignment, M$_S$ saturates at a lower applied field and H$_c$ is bigger if the fields are applied parallel to the c-axis versus perpendicular to this 'easy axis. M_r/M_s for isotropic unaligned samples is about half that of well orientated samples [12], and T$_C$ in oriented samples is likewise higher in the c-axis direction. The crystalline anisotropy, H$_A$, in A m^{-1}, determines the degree of this MCA, and the anisotropy constant K_1 indicates how difficult it is to shift the magnetisation out of that direction in the crystal lattice. The anisotropy constants K_1 and K_2 represent the energy required to change a magnetisation vector from its favoured low energy, or easy, orientation to a difficult, higher energy orientation. The total anisotropy energy for single hexagonal crystals is provided by the sum $\sum_k = K_o + K_1 \sin^2\phi + K_2 \sin^4\phi + ...$, where K_o is the energy to magnetise the easy axis and ϕ is the angle between the direction of magnetisation and the c-axis [40]. Higher order terms are rarely needed in uniaxial ferrites because K_o is low since the easy axis has a small energy orientation, and the subsequent order term is hardly used. For example, the anisotropy constants (measured in 10^2 Jm^{-3}) for BaM are $K_1>$ 3000 and $K_2 = 0$, making it an extremely magnetically hard material ideal for permanent magnets [12]. As a result, H_A is proportional to K_1 (which is positive), and the significant anisotropy in uniaxial ferrites is due to the contribution of the spin from the iron atom in the five-coordinate trigonal bipyramidal site [40].

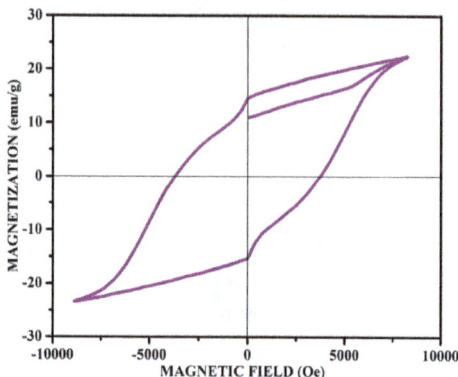

Figure 3. Hysteresis curve for the $SrFe_{12}O_{19}MNPs$ synthesised opting sol-gel method.

Although formed anisotropy opposes rotation inside the plane, it is a minor energy as that of H_A, which is high in all hexagonal ferrites. As a result, the magnetisation can freely spin within this plane, making ferroxplana ferrites poor permanent magnets. However, their little coercivity and high permeability cast them as an exceptional soft magnet for application in electrical circuits, with minute losses at high frequencies. At lower temperatures, the K_1 constant prevails, whereas at higher temperatures, the K_2 constant takes over, resulting in the observed variations in anisotropy of the ferroxplana ferrites [41]. The anisotropic partition of cations and unoccupied cation sites in ferroxplana CO_2 ferrites is thought to be the process, ensuing in a textured product with all of the particles' basal planes aligned [42]. Fig. 3 shows the hysteresis curve for the hexagonal $SrFe_{12}O_{19}$. The huge width of the loop indicates towards the high coercivity and therefore the permanent or hard magnetic nature of the strontium hexaferrite.

4.   Summary of hexagonal ferrites magnetic properties

Better sintered and denser ferrite has a higher magnetic moment per unit volume and thus, a higher M_s, making a composite magnet out of nanosized powders of high M_r magnets like a-iron and high coercivity magnets like $Nd_2Fe_{14}B$ can generate a material with high M_r and H_c, and this approach can also be utilised with ferrites. To boost coercivity, grain size should be below the domain size. Ferrites with magnetically aligned domains, on the other hand, will have a higher H_c, and orientated single crystal materials will approach the theoretical maximum values. The size and shape of the ferrite particles determine the magnetisation process under an applied field. The material is magnetically conditioned by rotating processes if it consists of plates under 5 μm; if it consists of grain exceeding 10 μm, wall movement dominates, with the essential grain size lowering at temperatures below room temperature [43].

5.   Strontium hexa ferrites (SrM)

SrM's magnetic characteristics are to some extent complex than those of BaM, with a value of 20.6μB. The Curie point is around 470°C [44], and the anisotropy constant is 3.5×10^6 erg cm^{-3}, resulting in a very high H_A in the c-axis of 1591 kAm^{-1} (20 kOe). The saturation magnetisation (M_s) of single crystal SrM has been recorded at various values ranging from 92.6 A m^2 kg^{-1} to 74.3 A m^2 kg^{-1}, and the greatest coercivity (H_c) is around 533 kA m^{-1}, Polycrystalline samples, on the other hand, hardly grasp these high levels. Because of the huge grain size of the early samples, the coercivity of polycrystalline SrM was first reported at relatively low values, and H_c was commonly quoted as 240 kA m^{-1}[45]. SrM with a diameter range of 0.5–50 μm has been reported to have a coercivity of 286 kA m^{-1}, and 0.1 μm specimens have been reported to have H_c = 517 kA m^{-1}[45]. SrM produced by the usual ceramic technique and then treated in nitrogen and hydrogen atmospheres before re-calcining in air yielded a material with a substantially greater coercivity of 400 kA m^{-1} and no loss of M_s in a grain size of less than 0.5 m. M_s = 101.3 A m^2 kg^{-1} at 5 K for SHF sintered at 1200°C for 4 hours [46].

A fine-grained (0.2 μm) sample was recently generated from co-precipitated salts, and when fired to 900–950°C, M_s and H_c were both 94 percent of the single crystal values, with M_s = 87 A m^2 kg^{-1} and H_c = 501 kA m^{-1}, respectively [34]. SrM has also been efficaciously formed from a sol gel precursor, comprised of mixed phases with α-Fe_2O_3 as an impurity product at 800 °C and 1000 °C, but single phase SrM at higher temperature [47].

The halide-based sol produced M_s = 63.3 kAm^{2}kg^{-1} and H_c = 453 kAm^{-1}, whereas the halide-free nitrate-based sol produced M_s = 65 kAm^{2}kg^{-1} and a significantly lower H_c = 440 kAm^{-1} [12], possibly because it formed SrM at lower temperature.

The low coercivity of conventional ceramic specimens can be considerably enhanced by lowering the grain size of the ferrite, just as they do with BaM. Milling can reduce size, although it is well known that milling reduces coercivity; in SHF samples 2–3 μm in diameter, H_c was lowered to only 159 kA m^{-1} after 24 hours of milling" [48]. After 80 hours of ball milling, the SrM had been subjected to such structural disorder that it had partially dissolved to α-Fe_2O_3, and super paramagnetic relaxation effects had drastically lowered H_c and M_s [49]. At 900 °C, SrM produced by chloride salt co-precipitation had a small particle size of 70 nm, resulting in extremely high coercivity in isotropic samples, with M_s = 71.8 Am^{2}kg^{-1}. The coercivity of a non-stoichiometric SrM with a slight iron shortage was found to be considerably greater, peaking at H_c = 541 kA m^{-1} with Fe:Ba ratio of 11.6, compared to a predicted maximum H_c of 597 kA m^{-1} [48]. Despite the strain created by the extended milling, SrM powder sized 0.5–50 μm was ball milled for 800°C/h to yield near-superparamagnetic sized powders of roughly 13 nm, which produced ferrite with a coercivity of 398 kA m^{-1} after being burned at 1000°C/ 4 h (with accompanying grain growth). The ferrite powder was even much smaller after vacuum milling, around 8 nm, and mixed with 3 nm magnetite crystals [45].

M_s = 70.7 Am^{2}kg^{-1} and H_c = 441 kAm^{-1} were obtained by milling a conventional ceramic sample to 0.8 μm, resulting in an oriented product with M_s = 70.7 A m^2 kg^{-1} and H_c = 441 kA m^{-1} [50]. About 75% oriented SrM was created by mixing the submicron SrM powder with 2–6% stearic acid in toluene, where stearate hydrophobic end is being connected to the solvent and the hydrophilic end to the SrM. Before sintering, the finely distributed ferrite was aligned and pushed in an external field. To obtain this degree of orientation, the product's coercivity had to be decreased first by milling the precursor to impart lattice strain, which dropped H_c from 438 kA m^{-1} to below 318 kA m^{-1} in samples burned to 1180 °C [50].

6.   Effect of substitution on magnetic properties of $SrFe_{12}O_{19}$ hexa hard ferrite

6.1   Substituted strontium hexaferrite (SrM)

SrM doped with up to 1% La_2O_3, with a concentration of 0.7 percent believed to be the best for boosting M_s and T_C while also raising the crystallisation temperature to 1200°C [51]. In RE: Ba ratios between 1:16 and 1:2, the substitution of RE metals into Sr site of

SrM was explored for La, Sm, and Nd. The Ms remained stable at roughly 65–67 Am^2kg^{-1} up to a ratio of 1:8, but it then declined with increased RE replacement, notably for Nd. Hc increased little and then fell beyond 1/8 for La substitution, Nd had a tiny positive effect, and simplified Hc significantly (by 25%), with the effect stabilising at a ratio of 1:4. All the RE ions appear to upsurge the isotropic after an initial reduction with 1:16 addition [52]. In a research paper, researchers discovered that they could make $Sr_{1-x}La_xM$ by hydrothermal synthesis when sintered at 1300°C, though they could not form the pure phase M ferrite for ratios of La:Sr> 1:8, and had unreacted impurities such as La_2O_3 and α-Fe_2O_3, emphasising the difficulty of making pure LaM [53]. Despite the fact that they were symmetrical-looking hexagons with no hint of DGG, they depict wide and thin hexagonal plates arise up to 5 m in diameter at 1:4 and 1:2, despite the fact that there were no noticeable alterations in microstructure up to a ratio of 1:8. The magnetic properties of ferrite powders reflected this shift in microstructure: for La:Sr ratios up to 1:8, there was essentially no change in Ms 64.5 A m^2kg^{-1} with formation temperature, as shown in their pure SrM sample. The Ms for a 1:4 ratio was significantly lower at 1000°C (56 Am^2kg^{-1}), and it only move toward the Ms of the further samples at 100 °C, where it stabilised at 63 Am^2kg^{-1}. The 1:2 ratio samples had a linear upsurge from 44.5 Am^2kg^{-1} at 1000°C to 62 A $m^2 kg^{-1}$ at 1300°C, illustrating the higher temperature required synthesising the samples with more La, as well as the fact that they never form the single-phase M ferrite, leaving non-ferromagnetic phases behind. Hc followed an analogous pattern, with pure SHF reaching a supreme Hc of 300 kAm^{-1} at 1100°C before declining as grain growth proceeded. The maximum values and peak Hc of the La substituted samples were similar. Although a little change in MCA, measurements of the anisotropy fields revealed that shape anisotropy was important, as small amounts of La substitution formed a microstructure that was favourable to coercivity, but for ratios >1:4 a less favourable microstructure of wide platy crystals was formed (resembling more the hexaplana ferrites). RE substitution in SrM had little influence on Ms in all cases, but raised Hc by up to 18% (Sm), 14 percent (Pr), 11 percent (Nd), and 5% (La) [54]. SHF was replaced with La^{3+} and Zn^{2+} to improve the magnetic characteristics, resulting in the compound $Sr_{1-x}La_xFe_{12-x}Zn_xO_{19}$. As the substituted ions became smaller than Sr^{2+} and Fe^{3+}, the lattice constant fell with x, and K_1 was reduced by 10% at x = 0.3. The grains were still only 0.8 μm after burning to 1200°C, resulting in a significant Hc of 374 kAm^{-1} and a 4 percent rise in Ms [45]. Sintered at 1215 °C and with x = 0.05–0.25, $Sr_{1-x}La_xFe_{12-x}Co_xO_{19}$ had a maximum Hc of 401 kAm^{-1} for x = 0.15 and a maximum Ms of 73 Am^2kg^{-1} for x = 0.18, while Tc declined linearly to a low of 430°C for x = 0.25 [26].

6.2   Magnetic properties

The effect of doping SrM with Sm^{3+}, Gd^{3+}, and Ho^{3+} ions on remnant magnetization (Mr), Ms, coercive field (H_c), and $(BH)_{max}$ is investigated in this section. The M-H curve of RE SrM at 300 K with a magnetic field of 10 kOe are shown in Fig. 4 [26] Due to the lack of Ms, the sample's Ms might be calculated using the law of approach to saturation (LAS) by eq. (1):

$$M = Ms\left(1 - \frac{A}{H} - \frac{C}{H^2}\right) + \chi H \tag{1}$$

where M denotes magnetization, A denotes inhomogeneity, c denotes high field susceptibility, H denotes applied field, and C denotes anisotropy. C can be calculated using eq. (2) for hexagonal ferrites:

$$C = \frac{8K_1^2}{105M_s^2} \tag{2}$$

The magneto crystalline anisotropy constant is K_1. Furthermore, as described in literature [55], the values of A/H and c in eq. (1) are slight for hexaferrite at suitably large magnetic fields. As a result, eq. (1) can be rewritten as eq. (3):

$$M = M_s\left(1 - \frac{8K_1^2}{105M_s^2H^2}\right) \tag{3}$$

Fig.5 shows the fitted curve for RE SrM 67.72 emu g^{-1} and 35.65 emu g^{-1} were the values calculated for the M_s and M_r of RE SrM respectively. These values are much higher than those reported in the literature for undoped SrM and some doped SrM. This can be described as follows: The enhanced Ms and Mr values of the RE SrM are due to the substitution of Fe^{3+} ions in the 2a (↑) and 12k (↑) octahedral positions with Sm^{3+}, Gd^{3+}, and Ho^{3+} ions. The theoretically calculated magnetic moments (μ_B) values of Gd^{3+} and Ho^{3+} ions, respectively, are determined to be 7 and 10 μ_B which are greater than the μ_B for Fe^{3+} ions [26].

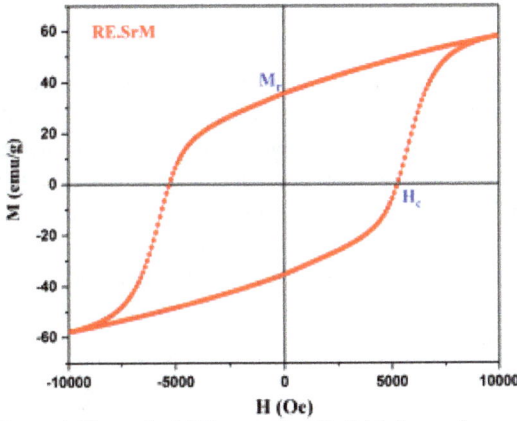

Figure 4. Shows the M-H curve for RE. SrM, from reference [26]

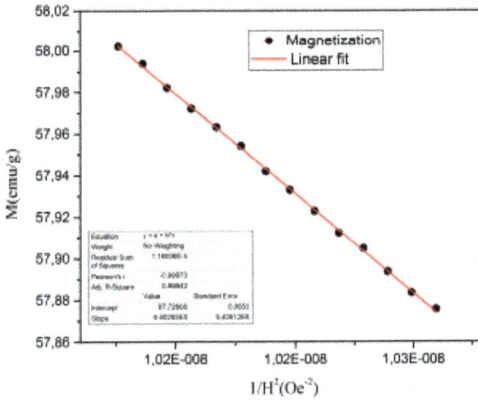

Figure 5. Shows M vs. 1/H2 curve for RE. SrM, from reference [26]

The MCA and exchange interactions can explain the improvement in coercive force (H_c=5257.63 Oe). Indeed, the anisotropy H_a is directly proportional to the H_c, and H_a is also proportional to the MCA constant K_1. As a result, H_c and K_1 are proportionate. H_a and K_1 have been determined to be $22.45*10^4$Oe and $7.602*10^6$ emu Oeg^{-1}, respectively. These results are better than previous studies on doped SrM (H_a=1.9053 104, K_1=0.5558 106 emu Oe g^{-1}) [56]. Furthermore, the "spin-orbital coupling" in RE ions is often higher than in 3d transition metal ions. As a result, replacing Fe^{3+} with Ho^{3+} and Sm^{3+} ions raise the coercivity value.

Because the 4f electrons of RE ions are shelled by the 5s and 5p electrons, the crystal field perturbation effects on them are weak. As a result, there are fewer quenching effects in RE ions than in 3d transition metal ions, resulting in strong spin–orbit interactions. The value of the coercive field of the RE, SM molecule is increased by doping with modest amounts of Gd^{3+}, Ho^{3+}, and Sm^{3+}. Indeed, the anisotropy of the Ho^{3+} and Sm^{3+} ions is large compared to that of a single ion, therefore they contribute to the anisotropy of RE. SrM [26]. Gd, on the other hand, has been shown to reinforce and contribute to anisotropy in the literature [57]. The *"magnetic super-exchange interactions"* between the 12k and $4f_2$ sites are seen schematically in Fig. 6.

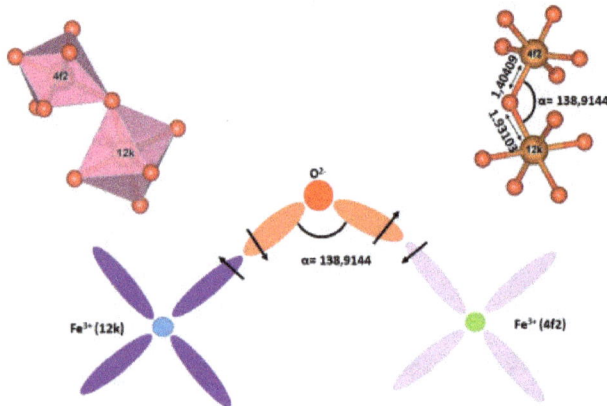

Figure 6. Schematic for the "magnetic super-exchange interactions" between the 12k and the $4f_2$ sites, from reference [26].

The M-H curve for $SrTb_{0.01}Tm_{0.01}Fe_{11.98}O_{19}/(CoFe_2O_4)_x$ (x =1.0, 1.5, 2.0, 2.5, and 3) hard/soft nano composites made through a one-pot sol–gel combustion technique is shown in Fig.7 (a-d). The most notable observation from the magnified views in Fig. 7 (c) and (d) is that the various *"hard/soft nanocomposites* created have great single-phase-like M–H curves. This means that the *"one-pot approach"* perfectly exchange-couples the hard $SrTb_{0.01}Tm_{0.01}Fe_{11.98}O_{19}$) and soft $(CoFe_2O_4)$ magnetic phases [58-59]. The H_c value increases as the soft phase content increases, as illustrated in Fig.7(c), which is unexpected. In the dearth of the $CoFe_2O_4$, direct coupling between $SrTb_{0.01}Tm_{0.01}Fe_{11.98}O_{19}$ hard ferrite grains take precedence, which has the potential to deviate magnetic moments from the easy axis due to *"magneto-static energy"* generated by dipolar contacts between hard ferrite grains. Due to the significant *"magneto-crystalline anisotropy"* of the hard ferrite grains, this phenomenon is not seen at modest reverse fields. However, as the reverse forces are intensified, the dipolar interactions among hard magnetic moments behave similarly to the applied fields, lowering the H_c value [60]. The direct coupling between hard grains is lessened when soft phase grains are included amid the hard phase grains. As a result, the deleterious impact of dipolar interactions is negligible.

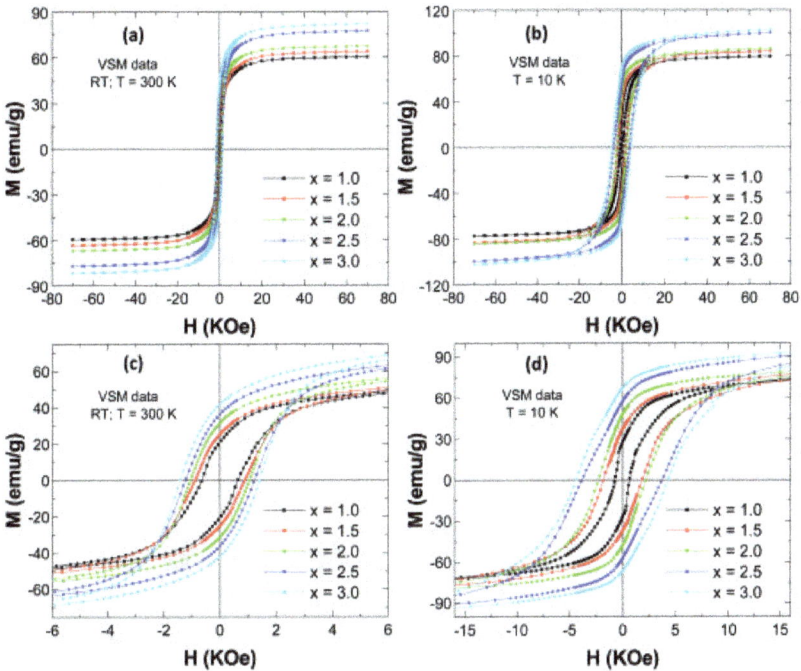

Figure 7. M-H loops at T = 300 K(a and c) and T = 10 K (b and d) of "SrTb$_{0.01}$Tm$_{0.01}$Fe$_{11.98}$O$_{19}$/(CoFe$_2$O$_4$)x (x = 1.0, 1.5, 2.0, 2.5, and 3.0) hard/soft nanocomposite from reference [58].

n the presence of soft ferrite grains, the hard/soft exchange-coupling interactions become more apparent, and the H_c value improved. The exchange-coupling contacts between hard-soft phases are thought to be exhausted in the presence of an excess soft phase, and *"dipolar interactions"* among soft grains become important, resulting in a lower H_c [60]. Furthermore, because the M_s forCoFe$_2$O$_4$ is larger than that of SrTb$_{0.01}$Tm$_{0.01}$Fe$_{11.98}$O$_{19}$, the M_s and M_r values increase as expected as the soft *"CoFe$_2$O$_4$"* phase content increases. Interactions between the interfacial spins of the soft and hard phases dictated the improvement in M_s and M_r values [61]

Fig. 8 shows the *switching field distribution* (SFD, i.e., dM/dH Vs. M curves) to further investigate the exchange behaviour [58]. A single peak in the dM/dH graphs indicates the finest *"exchange coupling"* between the soft-hard phases. If there is no *"exchange coupling"*, the SFD curves will show two separate peaks. Furthermore, when compared to the other samples, the nanocomposite with x = 3.0 showed the extreme peak intensity. This displays the sample's finest *"exchange coupling"*.

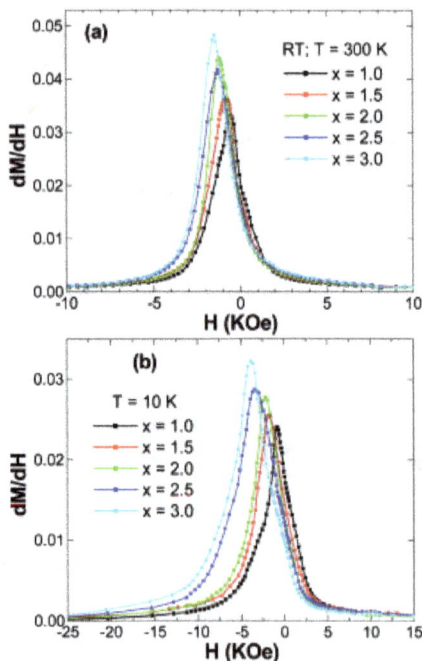

Figure 8. Shows the SFD plots at (a) T = 300 K and (b) T = 10 K of "SrTb$_{0.01}$Tm$_{0.01}$Fe$_{11.98}$O$_{19}$/(CoFe$_2$O$_4$)$_x$ (x = 1.0, 1.5, 2.0, 2.5, and 3.0) hard/soft nanocomposites", from reference [58].

7.   Magneto-dielectric properties

7.1   Dielectric properties

The "spark plasma sintering process" was used to create and compact the nanocomposite of $0.7BaTiO_{3-0.3}SrFe_{12}O_{19}$ [62]. In 50–300 K temperature range, the magnetic field and frequency dependency of its *"dielectric and magnetodielectric characteristics"* were investigated, with two different dielectric relaxation mechanisms observed. The dielectric tests for the BTO–SFO composite were carried out at 50–300 K temperature range keeping frequency between 0.1-100 kHz, as presented in Fig.9 Dielectric permittivity (ε`) vs frequency plot depict the distinct *"dielectric relaxation processes"* occurring. The dielectric polarizability of our composite, which contains a low ε`hexaferrite is likely to be impacted by a *"metal-like conductivity"* around ambient temperature [12, 62]. Fig. 9 shows that ε' has a substantial frequency dependence, especially at lower frequencies and high temperatures. At 220 K, the ε' values grow dramatically, especially at lower frequencies, where ε' can reach 3500 at f=100 Hz. The creation of *"Schottky-type diode"* at the interface is primarily responsible for such a large rise at very low frequencies, below 300 kHz. At intermediate frequencies, the apparent high value of ε' can be linked to "Maxwell–Wagner" and grain boundaries type *"interface relaxation processes"*, which are the dominating mechanisms at these frequencies" [63]. Two distinct plateaus can be found in the temperature range studied. There is a thermally triggered *"dielectric relaxation process"* between the two distinct plateaus at intermediate temperatures, which is owing to the occurrence of contemporaneous *"intrinsic/extrinsic contributions"* to ε' term. Fig. 9 depicts the evolution of ε' for selected frequencies as a function of temperature. There is a progressive rise of ε' in the lowest temperature range. A thin layer of $BiMnO_3$ exhibited similar behaviour, which was attributed to the material's usual ferroelectric nature [64].

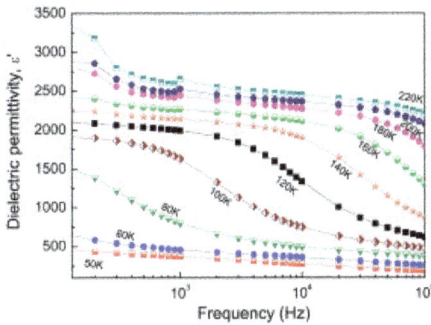

Figure 9. The frequency dependence of ε' in 50–220 K temperature range, from reference [64].

7.2   Magneto dielectric properties

Fig.10 (b) depicts the fluctuation of ε' with the external magnetic fields, given as *"MD=Δε'/ε'ₒ,"* where ε'_o represents the permittivity at zero-field. MD is magneto dielectric. At lower temperatures, there is a change in $\Delta\varepsilon'/\varepsilon'_o$. When fields up to 2.8 T were added to the *"dielectric tests"* done in the absence of external fields, an intrinsic positive *"MD effect"* of roughly 0.5–1% was identified at lesser temperatures, as shown in Fig. 10 (b). A closer look at the *"dielectric relaxation processes"* indicates that there is a discontinuity around the 1 kHz frequency (see Fig.9). These characteristics are temperature sensitive, appearing only above 160 K and increasing in amplitude until room temperature is achieved. The change of dielectric permittivity with temperature is seen in Figure 10 (a and b). The results were retrieved using ε'-magnetic field dependency measurements carried out at two fixed frequencies and a constant temperature. The effect of magnetic fields up to 5.6 Ton dielectric permittivity was investigated. In 0.555 kHz operating frequency, the observations emphasised in the two plots in Fig.11 undoubtedly indicate the significant effect of external fields on ε'. When the frequency is tuned to the resonance, the *"MD effect"* is clearly visible, as shown in Fig. 11(a). A continuous negative MD impact is detected in 150–202 K temperature range, with the maximum standards around 200 K [63].

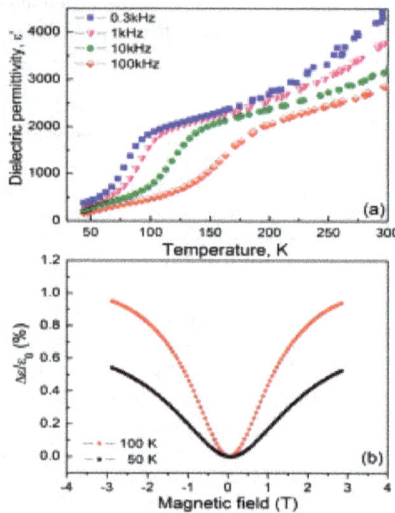

Figure 10. (a) Evolution of ε' with temperature measured at different frequencies. (b) Magnetodielectric effect observed at T=50 and 100 K., from reference [64].

According to the experimental findings, the $0.7BaTiO_3-0.3SrFe_{12}O_{19}$ composite has an MD impact across the entire temperature range tested, with an augmentation around ambient temperature [63]. The driving mechanism for such an effect is complex, but it may be guessed as a combination of an inhomogeneous medium and a magneto resistivity effect (MR), as seen in the inset of Fig 11 (a), resulting to a magnetic field dependent dielectric permittivity, as described in literature [65]. *As seen in* Fig. 11(b) *the MD effect decreases continuously at higher temperatures and becomes negligible around 270 K.*

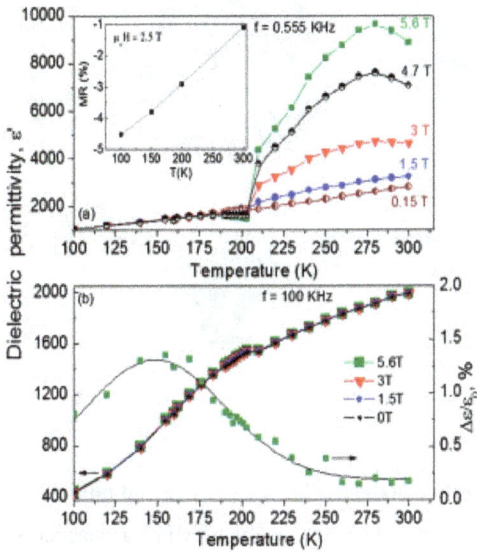

Figure 11. (a) Temperature dependency of ε'in various magnetic fields, recorded at 0.555 KHz "resonance frequency". The "MR effect" at a maximum magnetic field of 2.5 T is shown in the inset. (b) Dielectric permittivity temperature dependency when at 100 kHz frequency, from reference [64].

8.   Applications of hexagonal hard ferrites

Magnets are employed in a wide range of applications, including "actuators, motors, generators, sensors and transformers", as well as "data storage, mobile communications, transportation, security, defence, and aerospace, diagnostic devices, and electron beam focusing". Ferromagnetic alloys and metals, as well as ferrimagnetic ceramics, are the most commonly utilised magnetic materials. Hexagonal ferrites are the most widely used ceramic; some of their many applications are depicted in Fig.12.

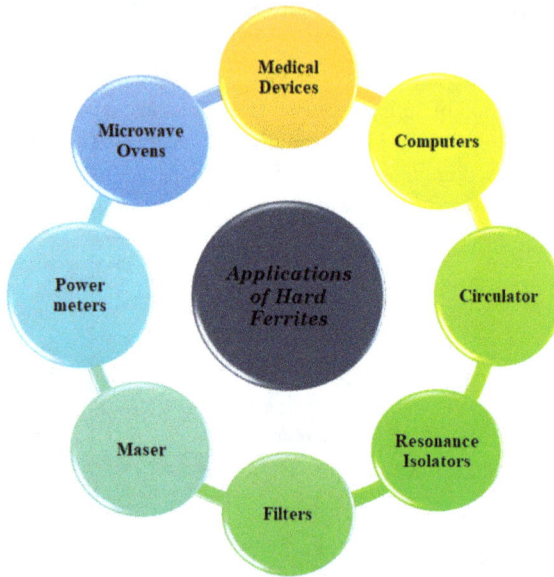

Figure 12. Applications of hard ferrites.

8.1 Advanced ceramic materials for microwave and millimeter wave engineering

Hexaferrites are magneto-dielectric materials with a magneto-plumbite crystalline structure. The general chemical formula for ferri-magnetic magneto-plumbite is $MeO_6Fe_2O_3$, where Me might be Ba^{2+}, Sr^{2+}, or Pb^{2+}. Al^{3+}, Ga^{3+}, Cr^{3+}, Sc^{3+}, or combinations of ions, such as Co^{2+} with Ti^{4+}, Zn^{2+}, etc., can partially replace ferric ions. Unlike spinel and garnets ferrites, hexagonal ferrites have significant core effective magnetic field which is related to magneto crystallographic anisotropy. Magnetically uniaxial hexaferrites, both monocrystalline and polycrystalline are the most commonly employed in practical applications. Because of the low Curie temperatures, the opportunities for researching planar hexaferrites are limited. For phenomenological descriptions of hexaferrite activity, the concept of crystallographic anisotropy, or simply anisotropy field, is often utilised.

Hexaferrites are an attractive substitute to cubic ferrites as "EMI suppressor components" since they have greater permeability, electrical resistivity and resonance frequencies, than cubic ferrites. Metallic ferromagnets, on the other hand, have a higher saturation magnetization but a sharp reduction in permeability as frequency rises due to eddy currents.

Materials Research Foundations **142** (2023) 93-120   https://doi.org/10.21741/9781644902318-4

They have, nevertheless, demonstrated a high potential for EMI suppressor devices when combined with hexaferrites. MNPs have a wide range of uses, including cell modification, detection, separation, and investigation, as well as the isolation of biologically active substances [66].

Some other applications of ferrites are listed below:

- **High-density write-once optical recording**

With blue wavelengths, thin films of defect ferrites can be employed as write-once, read-many media. Because these non-stoichiometric ferrites are metastable, a laser spot can convert them to corundum phases at moderate temperatures. The readout process is possible because the converted portions have different optical indices than the beginning ferrite film [67].

- **Magnetic sensors**

These are used to control temperature and can be constructed from ferrite with a precise Curie temperature. Ferrites have also been used to create position and rotational angle sensors (proximity switches). Shielding from magnetic fields A ferrite-based radar absorption paint has been created to make a submarine or aircraft undetectable to radar.

- **Pollution control**

Several Japanese systems use ferrite precursor precipitation to scavenge hazardous elements like mercury from waste streams. The ferrites formed as a result can be magnetically separated from the contaminant.

- **Ferrite electrodes**

Ferrites with the suitable conductivities have been utilised as electrodes in applications such as chromium plating because of their great corrosion resistance.

- **Power transformer and chokes:** HF Power supplies and lighting blasts
- **Inductors and tuned transformers:** Frequency selective circuits
- **Pulse and wideband transformers:** Matching devices

Conclusion

Hexaferrite remains the most practical material for applications, accounting for the great population of permanent magnets manufacturing today. They're fascinating materials with a wide range of applications. The SrM is of particular relevance, as it has piqued researchers notice because to its novel *"electromagnetic properties and potential applications"*. SrM is a favourite choice for permanent magnets used in ecologically friendly industrial applications, such as *"generator rotors for electric vehicles or wind energy"*, due to its low cost. The goal of this research is to offer fresh concepts for the development of MNPs appropriate for specific applications, as well as to explain how RE doping affects the magnetic properties of SrM.

When compared to Strontium-based composites, findings suggest that *"SFO–BTO composites"* are suitable materials for *"magnetic sensor devices"*. A favourable MD effect is reported at temperatures below 150 K, possibly due to induced strain mediated "spin–lattice coupling".

When looked at hard and soft "$SrTb_{0.01}Tm_{0.01}Fe_{11.98}O_{19}$/ $CoFe_2O_4$ nanocomposites" for various prepared hard/soft ferrite nanocomposites, a single peak in the SDF plots with no kinks were detected. The "hard ($SrTb_{0.01}Tm_{0.01}Fe_{11.98}O_{19}$)" and "soft ($CoFe_2O_4$)" phases were entirely *"exchange-coupled"* as a result of this. In all sectors of research, technology, and application, ferrites, particularly hard ferrites are naively assumed to be fully developed. Ferrite materials are now acknowledged as being more significant and essential for the advancement of electronics than they were previously and it is expected that ferrite manufacturing will expand year after year as their applications become more diverse. Reviewing the history of ferrite and precisely analysing its current status can tremendously aid future development.

Future outlook

The findings could be useful in the development of "future materials for microwave devices and recording media". As a generalized application potential, it's worth emphasizing that, given the encouraging results made in recent years in the development of ferrite MNPs technologies, even modest advances in ferrite magnets could lead to significant market substitution of RE magnets. Mild remanence improvements in motors and generators can result in large increases in power production. In this regard, collaborative efforts from both the magnetic and application sides are critical, as the definitive goal of replacement is to fabricate a competitive end device.

Permanent magnets are used in "low-carbon technologies" like renewable energy producers and electric vehicle engines and components. Over the next ten years, demand is likely to rise due to the urgent *"transition to a greener, more sustainable future"*. Because of the environmental impact of RE mining, if we rely solely on RE magnets, this change will be incomplete. As a result, ferrite-based motors for future mobility and renewable energy producers in the coming years become increasingly common.

References

[1] B. Want, B. H.Bhat, Magnetic and dielectric characteristics of Nd and Nd-Mg substituted strontium hexaferrite, Mod. Electron. Mater. 4, 21.(2018) https://doi.org/10.3897/j.moem.4.1.33273

[2] G. C.Papaefthymiou, Nanoparticle magnetism, Nano. Today. 4(2009) 438-447. https://doi.org/10.1016/j.nantod.2009.08.006

[3] K.Sato, when atoms move around, Nat.Mater. 8(2009)924-925. https://doi.org/10.1038/nmat2575

[4] G. Reiss, A. Hutten, Nat. Mater. 4 (2005)725-726. https://doi.org/10.1038/nmat1494

[5] J. H. Lee, J. T. Jang, J. S. Choi, S. H. Moon, S. H. Noh, J. W. Kim, J. W.Cheon, Exchange-coupled magnetic nanoparticles for efficient heat induction, Nat. Nanotechnol.6 (2011) 418-422. https://doi.org/10.1038/nnano.2011.95

[6] J. Zhang, J. Fu, F. Li, E. Xie, D. Xue, N. J.Mellors, Y. Peng, BaFe12O19 single-particle-chain nanofibers: preparation, characterization, formation principle, and magnetization reversal mechanism, Acs Nano. 6 (2012) 2273-2280. https://doi.org/10.1021/nn204342m

[7] Li, Q., Song, J., Saura-Múzquiz, M., Besenbacher, F., Christensen, M., M. Dong, Magnetic properties of strontium hexaferrite nanostructures measured with magnetic force microscopy, Sci. Rep. 6(2016) 1-7. https://doi.org/10.1038/srep25985

[8] P. E. Kazin, L. A.Trusov, D. D. Zaitsev, Y. D. Tretyakov, M. Jansen, Formation of submicron-sized SrFe12− xAlxO19 with very high coercivity. J.Magn.Magn.Mater. 320(2008) 1068-1072. https://doi.org/10.1016/j.jmmm.2007.10.020

[9] M. J. Iqbal, M. N. Ashiq, P. Hernández-Gómez, J. M. M. Muñoz, C. T. Cabrera, Influence of annealing temperature and doping rate on the magnetic properties of Zr-Mn substituted Sr-hexaferrite nanoparticles, J.Alloys Compd. 500(2010) 113-116. https://doi.org/10.1016/j.jallcom.2010.03.228

[10] H. Z.Wang, B.Yao, Y. Xu, Q.He, G. H.Wen, S. W.Long, L. L.Gao,Improvement of the coercivity of strontium hexaferrite induced by substitution of Al3+ ions for Fe3+ ions, J.Alloys Compd. 537(2012) 43-49. https://doi.org/10.1016/j.jallcom.2012.05.063

[11] D. Rawat, R. R Singh, Avant-grade magneto/fluorescent nanostructures for biomedical applications: Organized and comprehensive optical and magnetic evaluation, Nano-Struct. Nano-Objects. 26 (2021)100714. https://doi.org/10.1016/j.nanoso.2021.100714

[12] Pullar, R. C. Hexagonal ferrites: a review of the synthesis, properties and applications of hexaferrite ceramics, Prog. Mater. Sci. 57(2012)1191-1334. https://doi.org/10.1016/j.pmatsci.2012.04.001

[13] X.Meng, J. Mi, Q.Li, C.Bortolini, M.Dong, One dimensional BaxSr1− xEr yFe12−yO19fibers with magnetic crystalline nanoparticles, Mater. Res. Express. 1(2014)036106. https://doi.org/10.1088/2053-1591/1/3/036106

[14]S.Ounnunkad, Improving magnetic properties of barium hexaferrites by La or Pr substitution, Solid State Commun. 138(2006) 472-475. https://doi.org/10.1016/j.ssc.2006.03.020

[15]G.Nabiyouni, A.Ahmadi, D.Ghanbari, H. Halakouie, SrFe12O19 ferrites and hard magnetic PVA nanocomposite: investigation of magnetization, coecivity and remanence, J. Mater. Sci. Mater. Electron. J MATER SCI-MATER EL.27(2016) 4297-4306. https://doi.org/10.1007/s10854-016-4296-9

[16] F. Kools,A.Morel, R.Grössinger, J. M.Le Breton, P. Tenaud, (2002). LaCo- substituted ferrite magnets, a new class of high-grade ceramic magnets

intrinsic and microstructural aspects. J.Magn.Magn.Mater. 242(2002) 1270-1276. https://doi.org/10.1016/S0304-8853(01)00988-X

[17] Sharma, P., Verma, A., Sidhu, R. K., & Pandey, O. P. (2003). Influence of Nd3+ and Sm3+ substitution on the magnetic properties of strontium ferrite sintered magnets. J.Alloys. Compd. 361(2003)257-264. https://doi.org/10.1016/S0925-8388(03)00390-6

[18] M.Hassanpour, H. Safardoust-Hojaghan, M. Salavati-Niasari,Degradation of methylene blue and Rhodamine B as water pollutants via green synthesized Co3O4/ZnO nanocomposite, J. Mol. Liq. 229(2017) 293-299. https://doi.org/10.1016/j.molliq.2016.12.090

[19] M. Hassanpour, H. Safardoust-Hojaghan, M.Salavati-Niasari, Rapid and eco- friendly synthesis of NiO/ZnO nanocomposite and its application in decolorization of dye, J. Mater. Sci. Mater. Electron. J MATER SCI-MATER EL. 28(2017) 10830- 10837. https://doi.org/10.1007/s10854-017-6860-3

[20] L. A.Trusov, E. A.Gorbachev, V. A. Lebedev, A. E.Sleptsova, I. V.Roslyakov, E. S.Kozlyakova,P. E.Kazin, (2018). Ca-Al double-substituted strontium hexaferrites with giant coercivity, Chem. Comm. 54(2018) 479-482. https://doi.org/10.1039/C7CC08675J

[21] G. M.Rai, M. A.Iqbal, K. T.Kubra, Effect of Ho3+ substitutions on the structural and magnetic properties of BaFe12O19 hexaferrites, J.Alloys. Compd. 495(2010) 229-233. https://doi.org/10.1016/j.jallcom.2010.01.133

[22] S. Ounnunkad, P. Winotai, Properties of Cr-substituted M-type barium ferrites prepared by nitrate-citrate gel-autocombustion process, J.Magn. Magn. Mater. 301(2006) 292-300. https://doi.org/10.1016/j.jmmm.2005.07.003

[23] X. Liu, W. Zhong, S. Yang, Z.Yu, B.Gu, Y.Du, (2002). Structure and magnetic properties of La3+-substituted strontium hexaferrite particles prepared by sol-gel method, Physica status solidi (a). 193(2002) 314-319. https://doi.org/10.1002/1521-396X(200209)193:2<314::AID-PSSA314>3.0.CO;2-W

[24] J. F. Wang, C. B.Ponton,I. R.Harris, A study of Pr-substituted strontium hexaferrite by hydrothermal synthesis, J.Alloys Compd. 403(2005), 104-109. https://doi.org/10.1016/j.jallcom.2005.05.025

[25] J. F. Wang, C. B.Ponton, I. R. Harris, A study of Sm-substituted SrM magnets sintered using hydrothermally synthesised powders, J.Magn.Magn.Mater. 298(2006) 122-131. https://doi.org/10.1016/j.jmmm.2005.03.012

[26] M. Elansary, M. Belaiche, C. A.Ferdi, E.Iffer, I. Bsoul, New nanosized Gd-Ho-Sm doped M-type strontium hexaferrite for water treatment application: experimental and theoretical investigations, RSC Advan. 10(2020), 25239-25259. https://doi.org/10.1039/D0RA04722H

[27] L. M. Castelliz, K. M. Kim, P. S. Boucher, Preparation, stability range and high frequency permeability of some ferroxplana compounds, J. Can. Ceram. Soc.38 (1969) 57.

[28] K. Kamishima, N.Hosaka, K. Kakizaki,N.Hiratsuka, Crystallographic and magnetic properties of Cu2X, Co2X, and Ni2X hexaferrites, J. Appl. Phys. 109(2011) 013904. https://doi.org/10.1063/1.3527933

[29] S. I.Kuznetsova, E. P.Naiden, T. N.Stepanova, Topotactic reaction kinetics in the formation of the hexagonal ferrite Ba3Co2Fe24O41, Inorg. Mater., 24(1998), 856-859.

[30] J. Drobek,W. C. Bigelow, R. G. Wells, Electron microscopic studies of growth structures in hexagonal ferrites, J. Am. Ceram. Soc. 44(1961) 262-264. https://doi.org/10.1111/j.1151-2916.1961.tb15375.x

[31] F. M. M.Pereira, A. S. B.Sombra, A review on BaxSr1-XFe12O19 hexagonal ferrites for use in electronic devices,Solid State Phenom.202 (2013) 1-64. https://doi.org/10.4028/www.scientific.net/SSP.202.1

[32] E. P.Naiden, V. I.Itin, O. G.Terekhova, Mechanochemical modification of the phase diagrams of hexagonal oxide ferrimagnets, Tech. Phys. Lett. 29(2003), 889-891. https://doi.org/10.1134/1.1631354

[33] K. Haneda, C. Miyakawa, H. Kojima, (1974). Preparation of High-Coercivity BaFe12O19,J. Am. Ceram. Soc.57(1974) 354-357. https://doi.org/10.1111/j.1151-2916.1974.tb10921.x

[34] H. Yamamoto, H. Kumehara, R. Takeuchi, H. Nishio, Magnetic properties of Sr-M ferrite fine particles, Le Journal de Physique IV. 7(1997)535. https://doi.org/10.1051/jp4:19971219

[35] S. Okamoto, Structure of δ-FeOOH, J. Am. Ceram. Soc. 51(1968) 594-598. https://doi.org/10.1111/j.1151-2916.1968.tb13329.x

[36] E.Matijevic,Uniform colloidal barium ferrite particles. J. Colloid Interface Sci. 117(1987) 593-595. https://doi.org/10.1016/0021-9797(87)90426-7

[37] C. Sürig, D. Bonnenberg,K. A.Hempel, P. Karduck, H. J. Klaar, C. Sauer, Effects of variations in stoichiometry on M-type hexaferrites, Le Journal de Physique IV. 7(1997)315. https://doi.org/10.1051/jp4:19971124

[38] W.Zhong, W. Ding, N.Zhang, J. Hong, Q.Yan, Y. Du, Key step in synthesis of ultrafine BaFe12O19 by sol-gel technique, J.Magn. Magn. Mater. 168 (1997) 196-202. https://doi.org/10.1016/S0304-8853(96)00664-6

[39] R. C. Pullar, M. D. Taylor, A. K. Bhattacharya, A halide free route to the manufacture of microstructurally improved M ferrite (BaFe12O19 and SrFe12O19) fibres, J. Eur. Ceram. Soc. 22(2002) 2039-2045. https://doi.org/10.1016/S0955-2219(01)00518-0

[40] I. C.Heck, Magnetic Material and Its Applications, 1974.

[41] P.Muth, (1981). EP Wohlfarth (ed.). Ferromagnetic Materials, 2. (1981) 5210.

[42] C. H. Wilts, F. B. Humphrey, Magnetic anisotropy in flat ferromagnetic films: a review. J. Appl. Phys. 39 (1968) 1191-1196. https://doi.org/10.1063/1.1656219

[43] K. J. Sixtus, K. J. Kronenberg, R. K. Tenzer, Investigations on barium ferrite magnets, J. Appl. Phys. 27 (1956) 1051-1057. https://doi.org/10.1063/1.1722540

[44] B. T. Shirk,Ba2Fe6O11: A new metastable compound, Mater. Res. Bull. 5 (1970) 771-777. https://doi.org/10.1016/0025-5408(70)90091-7

[45] W. A.Kaczmarek, B.Idzikowski,K. H.Müller, (1998). XRD and VSM study of ball-milled SrFe12O19 powder, J.Magn.MagnMater. 177 (1998) 921-922. https://doi.org/10.1016/S0304-8853(97)00839-1

[46] N. Langhof, D. Seifert, M. Göbbels, J. Töpfer, Reinvestigation of the Fe-rich part of the pseudo-binary system SrO-Fe2O3, J. Solid State Chem. 182 (2009) 2409-2416. https://doi.org/10.1016/j.jssc.2009.05.039

[47] E. Otsuki, H. Matsuzawa, Magnetic properties of SrO.nFe2O3 powder synthesized by self-combustion process, In Journal de Physique IV Colloque 7(1997) 323. https://doi.org/10.1051/jp4:19971128

[48] F. M. M. Pereira, A. S. B.Sombra, A review on BaxSr1-XFe12O19 hexagonal ferrites for use in electronic devices, Solid State Phenomena.202, (2013) 1-64. https://doi.org/10.4028/www.scientific.net/SSP.202.1

[49] E. Wu, S. J. Campbell, W. A. Kaczmarek, A Mössbauer effect study of ball-milled strontium ferrite, J. Magn. Magn. Mater. 177 (1998) 255-256. https://doi.org/10.1016/S0304-8853(97)00910-4

[50] H. Taguchi, T. Takeishi, K. Suwa, K. Masuzawa, Y. Minachi,High energy ferrite magnets, Le Journal de Physique IV. 7 (1997) 311. https://doi.org/10.1051/jp4:19971122

[51] N.Dung, D. Minh, B.Cong, N.Chau, N.Phuc, The influence of La2O3 substitution on the structure and properties of Sr hexaferrite,In J. de. Phys. IV Colloque 7 (1997) 313. https://doi.org/10.1051/jp4:19971123

[52] R. Grossinger, M. Kupferling, J. C. Tellez Blanco, G. Wiesinger, M. Muller, G. Hilscher , et al, IEEE Trans Mag. 39 (2003) 2911. https://doi.org/10.1109/TMAG.2003.815745

[53] J. F. Wang, C. B.Ponton, R. Grössinger, I. R. Harris, A study of La-substituted strontium hexaferrite by hydrothermal synthesis, J.Alloys. Compd. 369 (2004) 170-177. https://doi.org/10.1016/j.jallcom.2003.09.097

[54] J. F. Wang, C. B. Ponton, I. R. Harris, A study of Pr-substituted strontium hexaferrite by hydrothermal synthesis, J. Alloys. Compd.403 (2005) 104-109. https://doi.org/10.1016/j.jallcom.2005.05.025

[55] Jr. Brown, W. F, The effect of dislocations on magnetization near saturation, Phys. Rev. 60 (1941) 139. https://doi.org/10.1103/PhysRev.60.139

[56] M. Z. Shoushtari, S. E. M. Ghahfarokhi, F. Ranjbar, Synthesis and Magnetic Properties of SrFe12-xCoxO19 (x= 0-2) Hexaferrite Nanoparticles, In Adv. Mater. Res.622 (2013) 925-929. https://doi.org/10.1007/s10948-014-2887-3

[57] B. A.Calhoun, M. J. Freiser,(1963). Anisotropy of gadolinium iron garnet, J. Appl. Phys. 34 (1963) 1140-1145. https://doi.org/10.1063/1.1729407

[58] N. A.Algarou, Y. Slimani, M. A.Almessiere,A.Baykal, Exchange-coupling behavior inSrTb0.01Tm0.01Fe11.98O19/(CoFe2O4)x hard/soft nanocomposites, New J. Chem.44(2020) 5800-5808. https://doi.org/10.1039/D0NJ00109K

[59] H. Yang, M. Liu, Y. Lin, G. Dong, L. Hu, Y. Zhang, J. Tan, Enhanced remanence and (BH) max of BaFe12O19/CoFe2O4 composite ceramics prepared by the microwave sintering method, Mater. Chem. Phys. 160(2015) 5-11. https://doi.org/10.1016/j.matchemphys.2015.04.032

[60] M. A. Radmanesh, S. S.Ebrahimi, (2012). Synthesis and magnetic properties of hard/softSrFe12O19/Ni0.7Zn0.3Fe2O4 nanocomposite magnets, J. Magn. Magn. Mater.324(2012) 3094-3098. https://doi.org/10.1016/j.jmmm.2012.05.008

[61] C. Pahwa,S.B.Narang, P. Sharma.P, Composition dependent magnetic and microwave properties of exchange-coupled hard/soft nanocomposite ferrite, J. Alloys Compd. 815(2020) 152391. https://doi.org/10.1016/j.jallcom.2019.152391

[62]M. Stingaciu, M.Reuvekamp, C. Tai, R. K. Kremer, M.Johnsson, The magnetodielectriceffect in BaTiO3-SrFe12O19 nanocomposites. J. Mater. Chem. C. 2(2) (2014) 325-330. https://doi.org/10.1039/C3TC31737D

[63] R. Schmidt, W. Eerenstein, T. Winiecki, F. D. Morrison, P. A. Midgley, Impedance spectroscopy of epitaxial multiferroic thin films,Phys. Rev. B: Condens. Matter Mater. Phys. 75 (2007) 245111. https://doi.org/10.1103/PhysRevB.75.245111

[64] M. Stingaciu, M.Reuvekamp, C. Tai, R. K. Kremer, M.Johnsson, The magnetodielectriceffect in BaTiO3-SrFe12O19 nanocomposites. J. Mater. Chem. C. 2(2) (2014) 325-330. https://doi.org/10.1039/C3TC31737D

[65] M. M. Parish, P. B. Littlewood, Magnetocapacitance in nonmagnetic composite media,Phys. Rev. Lett. 101 (2008) 166602. https://doi.org/10.1103/PhysRevLett.101.166602

[66] X. Li, J. Wei, K. E.Aifantis, Y. Fan, Q.Feng, F. Z.Cui, F. Watari, Current investigations into magnetic nanoparticles for biomedical applications, J. Biomed. Mater. Res. A. 104(5)(2016)1285-1296. https://doi.org/10.1002/jbm.a.35654

[67] L. Bouet, P. Tailhades, I. Pasquet, C. Bonningue, S. Le Brun, A. Rousset, Cation-deficient spinel ferrites: application for high-density write-once optical recording, Jpn. J. Appl. Phys. 38(3S) (1999) 1826. https://doi.org/10.1143/JJAP.38.1826

Chapter 5

Hard Ferrites for Permanent Magnets

Rohit Jasrotia[1,2*], Suman[3], Ankit Verma[4], Sachin Kumar Godara[5], Shubham Sharma[1], Kirti[1], Rahul Kalia[1,2], Pooja Puri[6], and Gagan Kumar[7]

[1]School of Physics and Materials Science, Shoolini University, Bajhol, Solan, H.P., India

[2]Himalayan Centre of Excellence in Nanotechnology, Shoolini University, Bajhol, Solan, H.P., India

[3]Department of Mathematics, School of Basic and Applied Sciences, Maharaja Agrasen University, Baddi, Himachal Pradesh, India

[4]Faculty of Science and Technology, ICFAI University, Himachal Pradesh, India

[5]Department of Apparel and Textile Technology, Guru Nanak Dev University, Amritsar

[6]Department of Chemistry, Bahra University, Waknaghat, Himachal Pradesh, India

[7]Department of Physics, Chandigarh University, Punjab, India

*rohitsinghjasrotia4444@gmail.com

Abstract

The hard ferrites called hexaferrites or hexagonal ferrites show excellent characteristics which makes them suitable for many potential applications. The magnetic characteristics of these hexaferrites must be tweaked to meet the needs of a certain potential application. This can be fulfilled by making few modifications within the crystal structure of hexagonal ferrites such as reduction of the size at the nanoscale, changes in the morphology, substitution of divalent as well as trivalent metal cations for increasing magnetization, and many more. In the current chapter, the requirement for excellent hexaferrite-based magnets, we evaluate the most promising ways for improving the performance of hexaferrite-based materials for utilization in permanent magnets, as well as their latest results. Then, we have provided a comprehensive study on the suggested modifications such as size reduction, substitution, shape and many more, which helps in the improvement of magnetic characteristics of hexagonal ferrites for their commercial use for permanent magnets application. Lastly, the concluding remarks have been presented.

Keywords

Hard Ferrites, Permanent Magnets, Size Reduction, Substitution

Contents

1.   Introduction

In a crystal, atomic moments interrelate with one other, resulting in a wide range of magnetic nature in various magnetic materials [1,2]. In general, these interactions can change the magnetic moments of individual atoms and even create new magnetic moments [3–5]. In general, magnetic materials may be divided into five groups: paramagnetic, diamagnetic, ferromagnetic, ferrimagnetic, and antiferromagnetic materials [6,7]. The magnetic field either produced by an electromagnet or permanent magnet (PM) is required to execute errands in both machines and devices. Humans have been using permanent magnets (PMs) for millennia, first in the form of lodestone [8] and later in the form of magnetized steel. The practical application of permanent magnets was restricted to the compass, and they were regarded as little more than a kind of magic trick by the majority of people. Magnets were first utilized in considerable quantities during the industrial revolution. Permanent magnet materials are now used in a variety of systems, including information storage, sensors, and electrical to mechanical transducer-based devices. The rise of technological evolution in the 20th century gave rise to the objective for the development of improving these historic magnetic materials [9]. Carbon steel magnets, alnico magnets, Sm-Co magnets, Nd-Fe-B magnets, and hexaferrite magnets/ferrite magnets/hard ferrite permanent magnets are the five kinds of permanent magnets for commercial application. This technical progress has been complicated by an increase in the tuning of the properties necessary for a hard magnet, most notably magnetic anisotropy. The permanent magnets show higher torque density, efficiency, and excellent performance of motors in comparison to electromagnets and therefore, due to these better characteristics of permanent magnets, they have many applications in the creation of machines and devices. The repeated use of permanent magnets does not deplete the energy stored in a permanent magnet; hence an ideal PM is a perfect power source. Permanent magnets also have the advantages of being simple to build and maintain, as well as having a low cost of production and operation. The advancement in the industrial and technological areas extensively increases the demand for the production of permanent magnets. In the year 2012, the automotive industry is the leading sector for the demand of PMs, but in 2019, the demand for PMs is increased by the larger expansion of the wind power industry. During the last few years, the Chinese government are working on various policies to limit the price of the generation of permanent magnets. In the last decades, the desire for higher performance magnets prompted extensive scientific research related to the creation of new approaches and the implementation of numerous scenarios aimed at increasing the magnetic characteristics of permanent magnet materials. Since their inception in the early 1950s, the number of published research papers on hexagonal ferrites has grown at an exponential rate. The anticipated alteration of a material's magnetic characteristics is, however, strongly reliant on the application for which it is created. For the manufacturing of permanent magnets used in high-power motors and drives, the highest possible remanent flux density and coercivity are required. Due to strong resistance to demagnetization by stray fields and the loss of recorded information, hexagonal ferrites called as hard ferrites would theoretically be excellent for magnetic data storage medium applications.In recent years, the hexaferrites known as hard ferrites, also known as M-type hexaferrites, have dominated the market for ferrite magnets, exceeding all other types of ferrite magnets. The

M-type hexaferrite-based materials are represented by the chemical formulas as ($BaFe_{12}O_{19}$ / $SrFe_{12}O_{19}$ / $PbFe_{12}$ / O_{19} / Other materials) [10–13]. In collaboration with Professor Kato, Takeshi Takei started the initial research for the first-time utilization of ferrites in the 20th century. In the first half of the 20th century, $BaFe_{12}O_{19}$ hexaferrite called BaM already existed. But, after the end of the second world war, under the supervision of Snoek, at the Philips Laboratories, BaM was observed to have a similar structure as that of the structure of PbM, as represented by $PbFe_{12}O_{19}$. In the year 1950s, all the major hexagonal ferrite-based phases were commercialized as described in Smit and Wijn's book but, furthermore, in 1952, BaM hexaferrite was fabricated for commercial use. Moreover, hexaferrites have an advantage such as low cost of starting precursors powders with 1.5 \$/kg whereas the starting precursors of NdFeB cost 60 \$/kg approximately. The Hexaferrite-based permanent magnets show superior higher values of resistivity along with higher chemical stability, which makes their utilization in mass applications. As a result of all these advantages, hexaferrite-based magnets are widely utilized for permeant magnets application. The ferrites show moderate values of saturation magnetization (M_s) and remanence magnetization (M_r) along with high coercivity (H_c). The magnetic hardness parameter (k) defines the utility of material for the permanent magnet's application, and also, the M-type hexaferrites have higher values of magnetic hardness, greater than 1. Therefore, the M-type hexaferrites are largely used for permanent magnets. Out of all hexaferrites, $BaFe_{12}O_{19}$ hexaferrite is the most utilized hexaferrite for the application of the permanent magnet as it shows a magnetic hardness of 1.34 approximately. In the market of permanent magnets, the higher grade hexaferrite magnets show a maximum remanence magnetization of 480 mega Tesla (mT) whereas, the Nd-Fe-B-based magnets show 1.3 Tesla (T) of remanence magnetization respectively [14]. In the last few years, the research in the area of hexaferrites has been mainly concentrating on hexaferrite-based nanomaterials, and therefore, this focused research of hexaferrites lead to many applications such as recording media, drug delivery, high frequency, microwave devices, wastewater remediation, etc. The information gained about nano magnetism, nanoparticle manufacturing, and nanoparticle processing could be vital for the revival of most conventional ferrite applications, such as permanent magnets. The main focus of this book chapter is to provide comprehensive knowledge about the utilization of hard ferrites called Hexaferrites or Hexagonal ferrites for the permanent magnet's applications are provided. The number of published Papers under the category of hexaferrites for permanent magnets has been provided in Figure 1.

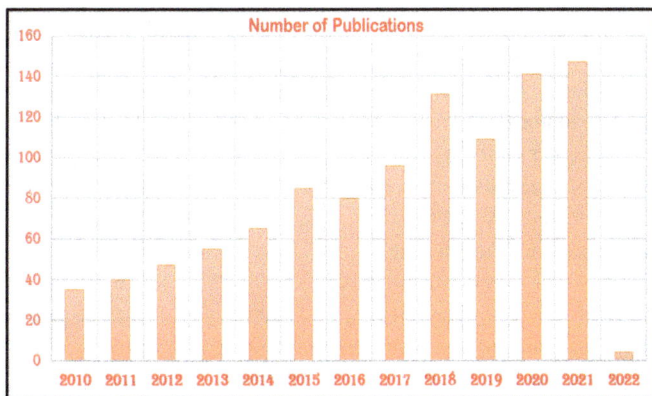

Figure 1. Number of Published Papers under the category of Hexaferrites for permanent magnets application (Taken from ScienceDirect database).

2.   Structure, properties, and characteristics of hard ferrites

The hexagonal ferrites belonging to a family of hard ferrites are based on the structure of magnetoplumbite with P63/mmc space geometry called $PbFe_{12}O_{19}$. For the indication of the planes, four indices are taken into consideration, out of which, the first three represent the within plane orientation at an angle of 120 degrees of unit vectors from each other, whereas the final one represents the c-axis respectively. The structure of hexaferrites is grounded based on the close-packed scheme of oxygen ions with the divalent and trivalent metal ions such as Barium, strontium, lead, lanthanum occupying the interstitial sites [15]. The schematic structure of M-type hexagonal ferrites is provided in figure 2.

Two building blocks were used to generate the crystal structure: the first is a two-layer block dubbed S-block (Spinel), and the second is a three-layer hexagonal close-packed of an oxygen ion doped in the middle hexagonal layer by larger divalent ions such as barium, strontium, lead, and lanthanum dubbed R-block. The S-block exhibits the same occupancy of A and B-sites (tetrahedral and octahedral) as the spinel structure, and so achieves the $Fe_6O_8^{2+}$ composition. Taking iron atoms at interstitial locations into account, the chemical composition of R-block is $BaFe_6O_{11}^{2-}$. The whole unit cell is constructed by sequentially arranging the blocks along the c-axis direction, following the sequence RSR*S, where * denotes a 180-degree rotation, resulting in a repetition distance of around 2.2–2.3 nm. The oxygen ion's radius is roughly four times that of the in-plane unit cell constant, which is 0.56 nm [16–19]. According to Wickoff positions 2a, 2b, $4f_1$, $4f_2$, and 12k, iron ions occupy five unique interstitial sites. Two statistically half-filled pseudo-four-tetragonal sites correspond to the five-fold coordinated (2b) geometry; additional interstitial sites include the A- and B-sites ($4f_1$ and 2a) of the S-block, the distorted octahedral sites ($4f_2$) of the R-block, and the distorted B-sites at the R-S block junction. The Neel model, which is used to explain the phenomena of antiferromagnetism, may be used to describe the magnetic structure of hexaferrites of the M type. Through superexchange interactions with the O^{2-} ions anions, the spin orientation of each Fe ion is coupled. The magnetic moment of ferric ions is exclusively determined by their spin. The B-sites in the S block point upward in the direction of overall magnetization, while the A-sites point downward in the opposite direction. The R-two block's B-sites are more correctly characterised as deformed A-sites, pointing downward. The bipyramidal site, in addition to the three octahedral sites, contributes two additional magnetic moments to the total magnetic moment. In contrast to ferrimagnetic oxides such as magnetite, the saturation magnetization (Ms) of M-type hexaferrites is not much less at room temperature, but it is very low in comparison to the Ms of the majority of materials used in permanent magnet applications.

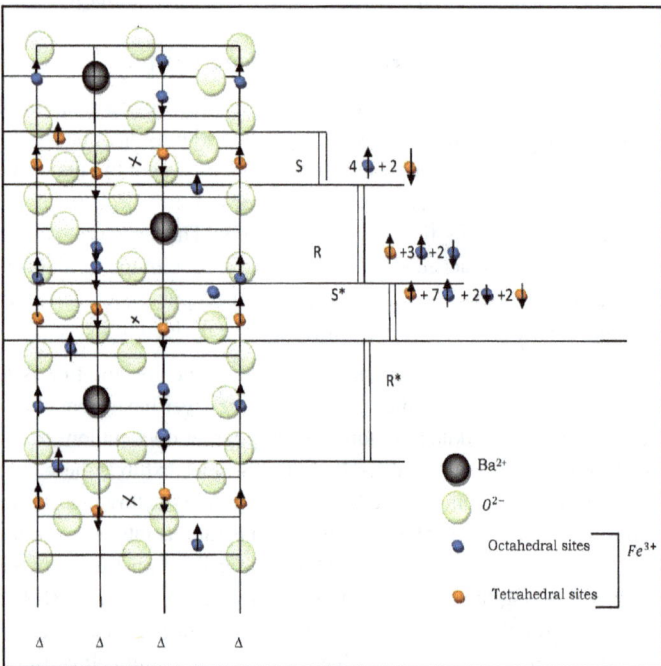

Figure 2. Schematic structure of the M-type hexaferrite.

3.   Generation of commerciallyavailable permanent magnets

The loadstone utilized in ancient compasses created in China over 3000 years ago is the oldest recorded material for permanent magnets. This substance, which has been around 5000 years, is mostly made up of the mineral Fe_3O_4 (magnetite). Magnetite is a soft magnetic material that maybe demagnetized by a weak external reversal field. As a result, this substance is ineffective as apermanent magnet. The demand for permanent magnets with improved mechanical and magnetic characteristics prompted an intensive study into the creation of permanent magnets with excellent performance. As a result of these efforts, today, permanent magnets with excellent performance have been discovered. The most significant commercial permanent magnets, as well as the first magnet called carbon steel manufactured for practical purposes, are explained below. The schematic representation of commercially available permanent magnets is provided in figure 3.

3.1   Carbon steel magnets

The manufacture of quench hardened iron-carbon alloys (sword steel) was the result of William Gilbert's (1540 – 1603) experimental work on magnetism, which was reported in his book On the Magnet in the 1600s. More than 150 years after Gilbert's invention, attempts to manufacture iron magnets resulted in improvements in the production of permanent magnets for potential; applications, but no significant breakthroughs in the sector. Steel magnets consisting of magnetic steel strips were used to raise iron items in the seventeenth century. At the time, the strips were magnetized by rubbing them with a loadstone. However, it took Hans Christian Oersted's (1775 – 1851), a significant discovery in 1820 that a magnetic field may be produced by an electric current, and the subsequent invention of the first electromagnet in 1825, for real research on magnetic materials to begin [7].

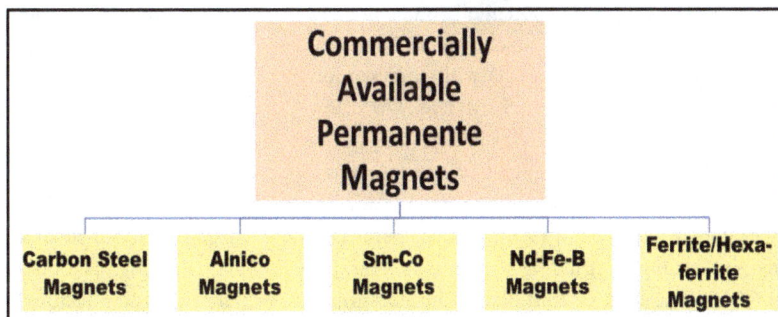

Figure 3. Five types of commercially available permanent magnets.

Magnetite has a saturation magnetism that is superior to carbon steel magnets. Carbon steel's low coercivity, which can be demagnetized by a small reversal field of only 50 Oe, is an evident disadvantage for operation in the presence of stray fields. Another disadvantage is the low energy product of <0.25 MGOe (2 kJ/m^3). After 1880, it was discovered that alloying steel with other metals improved its magnetic qualities, and by 1885, steel with 5% tungsten was being produced and utilized for magnets. During World War I, this was replaced by the less expensive chromium steel. Tungsten and chromium steels, on the other hand, had a coercivity of less than 100 Oe until 1900, and their energy product was less than 0.3 MGOe (2.4 kJ/m^3). Efforts to increase the magnetic characteristics of steel magnets led to the finding that adding cobalt to the alloy might triple the coercivity. In 1917, Kotaro Honda filed three patents for the discovery of novel carbon steels with improved magnetic characteristics, and in 1920, he patented the invention of carbon steel with 35% cobalt [8]. Carbon steels containing 30-40 percent cobalt, as well as tungsten and chromium, have been claimed to have an increased energy product of up to 1 MGOe (8 kJ/m^3) and the highest coercivity of any magnetic material (230 Oe). Steel magnets, on the other hand, had significantly poorer magnetic characteristics than those produced a decade later. Furthermore, due to their high percentage of the pricey cobalt element, the best steel magnets are expensive. As a result, commercial production of these magnets ceased, and they were replaced by modern, high-performance magnets.

3.2 Alcino magnets

Tokuhichi Mishima's patent in 1931 for the first magnet alloy based on Iron, Nickel, and Aluminium, signified a shift in the scientific community's focus towards the creation and manufacture of a new type of material for high-performance magnets. The new $Al_{12}Ni_{30}Fe_{58}$ alloy exhibited a coercivity of 400 Oe, almost double that of the finest steel magnet, according to Mishima. The insertion of new elements, primarily Cobalt, as well as minor amounts of copper and potentially other elements, increased the magnetic properties of the Mishima magnet. The development of these materials over a three-decade period resulted in the creation of magnets with coercive forces H_{cB} ~1900 Oe (150 kA/m) and energy products surpassing 9.5 MGOe (76 kJ/m^3), enhancing permanent magnet application in the machine industry. Alnico magnets also have a strong remanence (maximum B_r = 1.33 T) and a high Curie temperature (Tc = 850°C), making them suitable for use in high-temperature regimes ($\geq$ 500°C). Alnico alloys are brittle and hard, thus they can't be cold-worked. As a result, alnico magnets manufacturing is confined to either melting and casting or grinding to a fine powder, pressing, and sintering. The magnet's final polishing process is confined to surface grinding. Sintered magnets feature finer grains than cast magnets, are mechanically stronger, and have a better surface polish. Sintered magnets, on the other hand, have lower magnetic characteristics than cast magnets. The magnetic characteristics of as-cast or as-sintered magnets can be improved by using particular heat and magnetic field treatments to homogenize the solid solution and increase remanence and coercivity. The high remanence and high-temperature performance of Alnico magnets are their main advantages. Commercially, available alnico products with an operating temperature of 500° C have the following magnetic properties: H_{cB} = 150 kA/m, B_r = 1.12 T, and $(BH)_{max}$ = 92 kJ/m^3. However, due to the need for increased coercivity and energy products for device downsizing, as well as poor mechanical qualities (brittleness and hardness) and a relatively high market price for cobalt, the production of such magnets has been constrained [20].

3.3   Sm-Co magnets

The development of the (RE-TM) alloys series resulted in magnets with dramatically better magnetic characteristics. The discovery of YCo$_5$'s strong magnetocrystalline anisotropy in 1966 sparked interest in developing Re-Co intermetallic compounds for permanent applications [21]. Karl Strnat discovered SmCo$_5$ in the mid-1960s, and its energy product of 143 kJ/m^3 was significantly superior to that of other permanent magnets at the time. In 1970, the product debuted as the first generation of high-performance (1-5) RE magnets. The coercivity of this molecule was found to be substantially higher than that of YCo$_5$ and other RE-Co derivatives [22]. Although this intermetallic has a Curie temperature of 750°C, the behavior of the coercivity with rising temperature limits its application to temperatures below 250°C. These working temperatures are high but not as high as Alnico magnets. Furthermore, the high cost of this magnet is a drawback due to the large content of the pricing of Sm metal (34 wt.%). Different processing approaches, as well as RE and TM substitution scenarios, were used to alter the characteristics of the 1-5 magnets. Partial substitution of Samarium and protium leads to a minor cost decrease and an improvement in the energy product, but it reduces the product's coercivity. In the early 1970s, Sm-Pr-Co$_5$ magnets with an energy product of around 200 kJ/m^3 were manufactured, only a few years after sintered SmCo$_5$ magnets with an energy product of slightly more than 160 kJ/m^3 were produced. By adding trace quantities of copper, zirconium, and iron, the energy product increased significantly to 239 kJ/m^3. The cost of replacing Sm with Ce is lower, but the energy product and curie temperature of the product is lower. B_r = 0.50 – 1.03 T, H_{cj} = 398-1990 kA/m, and $(BH)_{max}$ = 47.8 – 207 kJ/m^3 were reported for complex combinations of Sm(Ce, Pr)-Co(Fe, Cu) "1-5" magnets [23]. The introduction of heavier rare earth metals (such as Gadolinium) improved performance at higher temperatures, but did so at the expense of increased magnet cost, reduced remanence, and lowered the energy product to 72–111 kJ/m^3. Copper substitution for Cobalt can also cut the magnet cost, allowing for the insertion of iron into the product, although this technique has the drawback of lowering the remanence and energy product. B_r = 0.61 T, H_{cB} = 279 kA/m, and $(BH)_{max}$

= 55 kJ/m^3 were reported for a 1-5 magnet with an optimal composition of 31 wt.% Samarium, 46 wt.% Cobalt, and 23 wt.% Copper [24]. Several metal substitutes for Sm and Co can thus be used to modify the properties of the 1-5 magnets, but it appears challenging to achieve benefits in all essential parameters and product cost. Commercially available 1-5 magnets with top working temperatures of 200–350°C and the following magnetic properties: B_r = 0.59–1.0 T, H_{cJ}= 358–2000 kA/m, H_{cB} = 358–770 kA/m, and $(BH)_{max}$ = 68–190 kJ/m^3 are currently available. SmCo$_5$ magnets have the benefit of being charged (magnetized) in a field much smaller than their inherent coercivity, which reduces production costs. This is owing to the magnet's magnetization-demagnetization and nucleation-controlled coercivity behavior. Domain wall motion is used to achieve the initial magnetization, which necessitates relatively tiny applied fields. The demagnetization curve is nearly flat until a reverse field close to intrinsic coercivity is applied, at which point the magnetization rapidly declines. The nucleation of reverse domains causes magnetization reversal in SmCo$_5$, which necessitates a strong field. The nucleation field is sufficiently powerful to induce the grains' domain walls to migrate in the other direction and approach saturation [25]. The second generation of RE magnets is the 2–17 family, which was introduced in 1980 with the nominal commercial composition Sm(Co, Fe, Cu, Zr)$_{7.4}$. Because this magnet has a lower Sm content than the 1-5 family, it is projected to be less expensive. However, the greater processing cost offsets the lower raw material cost, keeping this magnet's market price near the top of the PM Price list. Although these magnets are marketed as "2-17", the commercial products of this family are thought to be made up of 2-17 phase regions separated by 1-5 phase bands. This magnet's process for obtaining high coercivity differs from that of SmCo$_5$; it is caused by domain wall pinning rather than domain nucleation, as in SmCo$_5$. The Br of the 2–17 magnet family is higher than that of the 1-5, resulting in a higher energy product (159–239 kJ/m^3). Continuous research efforts resulted in the development of a range of 2-17 magnets with increased coercivity, remanence, and energy products. In 1990, Liu & Ray reported the creation of a magnet containing light RE metal with the nominal composition Sm$_{0.8}$(Ce$_{0.2}$Pr$_{0.4}$Nd$_{0.4}$)$_{0.2}$(Co$_{0.633}$Fe$_{0.286}$Cu$_{0.061}$Zr$_{0.020}$)$_{7.59}$, with the following magnetic properties: B_r = 1.157 T, H_{cJ} = 1234 kA/m, H_{cB} = 822 kA/m, and $(BH)_{max}$ = 239 kJ/m^3. However, utilizing a greater Fe content in Sm-Co compound with nominal composition Sm(Co$_{0.612}$Fe$_{0.316}$Cu$_{0.052}$Zr$_{0.020}$)$_{7.73}$ resulted in the maximum energy product of 271 kJ/m^3. Furthermore, it was discovered that the intrinsic coercivity of the compound Sm(Co$_{bal}$Fe$_{0.06}$Cu$_x$Zr$_{0.03}$)$_z$ is sensitive to x and z values, with z being more susceptible. At z = 5.8, the intrinsic coercivity was 438 kA/m for x = 0.1 and 0.12, and grew substantially to 3184 kA/m at z = 7.2. The intrinsic coercivity increased from 119 kA/m at z = 5.8 to 1880 kA/m at z = 7.2 at a reduced Cu concentration of x = 0.8. Commercially available Sm$_2$Co$_{17}$ magnets with high magnetic characteristics and working temperatures ranging from 200 to 350°C were developed. Commercial grades of 2–17 magnets with magnetic characteristics $B_r = 0.93 – 1.15$ T, $H_{cJ} = 636 – 1990$ kA/m, and $H_{cB} = 557 – 845$ kA/m, and $(BH)_{max} = 160\text{-}255$ KJ/m^3 are already available. The performance of Sm-Co magnets degrades over 300° C, limiting their use in high-temperature applications. The goal, therefore, became to develop Sm-Co magnets with greater working temperatures. At high temperatures, the intrinsic coercivity of $Sm(Co, Fe, Cu, Zr)_z$ ($z = 6.7\text{-}9.1$) was examined. Liu et al. created magnets with the greatest ever known coercivity of more than 796 kA/m at 500° C in that research. $Sm(Co_{6.1}Cu_{0.6}Ti_{0.3})$ with a coercivity of 685 kA/m at 500° C was also successfully produced in the lab. This chemical was described as a two-phase combination of Sm-Co phases 1–5 and 2–17. This high-temperature magnet, however, has a lower residual flux than Alnico and a lower energy product (31.8 kJ/m^3) than other commercially made Sm-Co magnets. The insertion of iron improved the high-temperature magnetic characteristics marginally, with $Sm(Co_{6.0}Fe_{0.4}Cu_{0.6}Ti_{0.3})$ having an intrinsic coercivity of 764 kA/m at 500° C, and the energy product and magnetization improved [20].

Powder-metallurgical techniques are used to make commercial $SmCo_5$ permanent magnets from alloys produced using one of the following processes:

- Induction melting and crucible casting of metallic components.

- Reduction-diffusion calcio-thermic (R-D).

The as-cast alloy is crushed and milled in the first step to produce micron-sized particles. Anisotropic magnets are made by aligning the particles in a magnetic field and crushing them in a die. After that, the compacts are sintered at a temperature of around 1000°C. Before compaction, a little amount of powder containing excessive levels of Sm is occasionally added to the magnet powder. The liquid-phase sintering procedure results in higher density and improved magnetic characteristics since this powder melts at the sintering temperature. The mean grain size of the sintered alloy is orders of magnitude larger than the critical single-domain particle size, and the alloy's multi-domain structure enables charging. Due to the sintered alloy's multi-domain structure, it is possible to charge the magnet with a field that is far smaller than the magnet coercivity, as previously stated. The mean grain size of the sintered alloy is an order of magnitude greater than the critical single-domain particle size, and the alloy's multi-domain structure allows it to charge the magnet with a field that is far smaller than its coercivity. Cutting or grinding into the final required shape, as well as re-magnetizing, if necessary, completes the PM manufacturing process.The RE and other transition metals are employed in oxide powder form in the second phase (also known as the KOR process), whilst Cobalt and Iron are introduced as metal powders. By causing simultaneous diffusion of the powder ingredients, the calcio-thermic reduction of oxides with calcium results in a spongy alloy powder. This method reduces the cost of producing RE magnets by eliminating many processes. To obtain final goods, the powders are processed in the same manner as in the first procedure. Due to the high chemical reactivity of RE and alloy powders, extra vigilance and the use of a vacuum or an inert environment are required during processing. The magnetic powders could also be used to make bonded magnets, which are made by molding or pressing appropriate combinations of the magnetic material with a binder (plastic, rubber, or a low-melting-point substance). The curing procedure is completed with subsequent heat treatment. Finally, due to the ductility of the material, magnets of any shape are simple to make. RE-bonded magnets have been manufactured since the mid-1970s, but due to their inferior magnetic characteristics compared to sintered magnets, they have received little attention. Later in the 1980s, Japanese efforts were successful in establishing boded RE-magnets as a popular magnet. Early $SmCo_5$ bonded magnets had energy products of 3–10 MGOe (24–80 kJ/m^3), which were comparable to or better than the energy products of today's top alnico or ferrite magnets. The technique for making 2–17 magnets is largely the same as for making $SmCo_5$ magnets, with the exception that additional heat treatment is required to get the microstructure required for optimal magnetic characteristics. For the manufacturing of the 2–17 magnets, the reduction-diffusion method was effectively adjusted [20].

3.4   Nd-Fe-B magnets

Due to the paucity of Sm and the high market price of Sm and Co metals, scientists have been working hard to develop less expensive materials with improved magnetic characteristics. John Croat's group at General Motors reported the fabrication of RE-Fe binary alloys with promising magnetic characteristics for permanent applications in a series of articles in the early 1980s [25,26]. The alloy with the greatest intrinsic coercivity among the Re-Fe alloys was determined to be $Nd_{0.4}Fe_{0.6}$, which had a coercivity of 593 kA/m [26]. At the same time, Koon and colleagues published the results of their research on (Fe, B)$_{0.9}Tb_{0.05}La_{0.05}$ alloys [27,28] claiming that the alloy $(Fe_{0.82}B_{0.18})_{0.9}Tb_{0.05}La_{0.05}$ has an inherent coercivity of 716 kA/m and remanence of 0.5 T. The findings of these tests lay the groundwork for the development of new Nd-Fe-B magnets. In April 1983, Hadjipanayis and associates at the time at Kollmorgen Company in Radford, Virginia) boosted the speed of research devoted to the development of cheaper, more efficient magnets. Pr-Fe-B-Si magnetic material generated by melt spinning and annealing was one of their fascinating products, and an article describing the experimental results was submitted to Applied Physics Letters on June 23rd, 1983, and accepted for publication on August 1st of the same year. The compound $Fe_{76}Pr_{16}B_5Si_3$ has a maximum energy product of 95.5 kJ/m^3, making it an excellent Co-free PM material [29]. Sumitomo Special Metals

Company of Osaka, Japan, stated in June 1983 that it had produced a novel magnet using a powder metallurgical process with a record energy product. Sagawa and colleagues examined the structural and magnetic characteristics of $Nd_xB_yFe_{(100-x-y)}$ intermetallic (with $x = 13$–19 and $y = 4$–17) [30]. The RE (Nd) concentration of these intermetallic was marginally higher than that of Koon's compound $((Tb, La)_{10}Fe_{74}B_{16})$ [22], but much lower than that of Croat's Nd-Fe binaries [26,31]. The intrinsic coercivity of $Nd_xB_8Fe_{92-x}$ was shown to improve up to $x = 15$, then somewhat improve at increasing Nd content, whereas B_r and BH_{max} were at their best in this range $x = 14 - 15$. The intrinsic coercivity of $Nd_{15}B_yFe_{85-y}$, on the other hand, improved up to $y = 8$, and only minimally with higher B-content, whereas Br and BH_{max} revealed ideal values at $y = 6$–8. $Nd_{15}B_8Fe_{77}$ was discovered to crystallize into a tetragonal phase with $a = 0.880$ nm and $c = 1.221$ nm, increased magnetic parameters of $B_r = 1.23$ T, $H_{cJ} = 960$ kA/m, $H_{cB} = 880$ kA/m, and a record energy product of $(BH)_{max} = 290$ kJ/m^3 [30]. The intrinsic coercivity of this compound increased when the maximum magnetising field strength was increased, reaching $H_{cJ} = 850$ kA/m when a magnetising field of 600 kA/m was applied, and then the squareness improved, and H_{cJ} reached the optimal value when a magnetising field of 1200 kA/m was applied. On the opposite side of the Pacific, various American organisations presented their discoveries on the unique RE-Fe-B magnetic materials at the 29th Annual Conference on Magnetism and Magnetic Materials in Pittsburgh in November 1983. Koon and Das addressed the manufacture of Pr-Fe-B and Nd-Fe-B magnets, as well as their annealed melt-spun ribbons that produced an energy product of 103 kJ/m^3. Croat, a subsidiary of General Motors, announced the manufacture of Pr-Fe-B and Nd-Fe-B magnets with an energy product of 120 kJ/m^3 and the discovery of a way to double this value. Croat et al. published an article in Applied Physics Letters in 1984 describing the fabrication of an Nd-Fe-B magnet with an energy product of 112 kJ/m^3, claiming it to be the highest recorded energy product for a light RE-Fe material. Furthermore, Croat's group's results on a series of Nd-Fe-B magnets appeared in the same volume of the Journal of Applied Physics in 1984, immediately before Sagawa's group's article. Croat et al. published their findings on a variety of (Nd, RE)-Fe-B compounds synthesized by melt spinning under various experimental settings in the latter publication. The $Nd_{0.135}(Fe_{0.935}B_{0.065})_{0.865}$ composition had an improved intrinsic coercivity of around 15 KOe (1194 kA/m), a remanence of about 8 kG (= 0.64 T), a maximum energy product of 14 MGOe (111 kJ/m3), and a remanence of about 8 kG (= 0.64 T). The Japanese product outperforms the American product in terms of remanence and energy output, whereas the American product outperforms the Japanese product in terms of magnetic hardness. When the magnetic properties of the compounds investigated by the Japanese and American groups were compared, it became clear that the small compositional differences between the compounds investigated by the Japanese and American groups could not account for the observed significant differences in magnetic properties. As a consequence, the observed discrepancies may suggest that the magnetic properties of Nd-Fe-B magnets are process- and sample-dependent. When compared to Sm-Co magnets, Nd-Fe-B magnets have a Curie temperature of T_C 310°C, which restricts their performance at high temperatures. NdFeB magnets with a high flux density and a high energy product with B_r = 1.45 T, H_{cJ} = 1114 kA/m, H_{cB} = 1080 kA/m, and $(BH)_{max}$ = 414 kJ/m^3 with a working temperature of up to 100°C are now commercially available. Although some of the first Nd-Fe-B magnets had high flux density and energy products, their non-linear B-H curves and instability at high temperatures made them unsuitable for motor applications. As a consequence, significant effort has been undertaken to enhance the performance of NdFeB magnets at elevated temperatures. While the B-H curve of an Nd-Fe-B magnet had a knee at ambient temperature, partial substitution of Nd by Dy ($Nd_{10}Dy_4Fe_{80}B_6$) improved the coercivity and maintained a linear B-H behavior up to 175°C, thus increasing the magnet's characteristics for high-temperature applications. However, this substitution reduces the remanent flux density and energy product while also increasing the magnet production cost. Commercial grades of high-coercivity NdFeB-based magnets with the following magnetic characteristics are now available, functioning at higher temperatures (up to 220°C): H_{cJ} = 2662 kA/m, H_{cB} = 971 kA/m, and $(BH)_{max}$ = 310 kJ/m^3. B_r = 1.27 T, H_{cJ} = 2662 kA/m, H_{cB} = 971 kA/m, and $(BH)_{max}$ = 310 kJ/m^3. At comparable operating temperatures, these magnetic characteristics outperform Sm-Co magnets. This appears to be the type of PMs the world has been waiting for: a high-performance, Co-free magnet, especially given the unanticipated variations in Co-price as a result of increased market demand and political unrest in the supplying regions. As a result, Nd-Fe-B magnet production increased exponentially from 1997 to 2002, and the exponential rate of rising was accelerated from 2003 to 2006. This strategic PM's market sales grew from $1.142 billion in 2003 to $3.436 billion in 2008. In theory, Nd-Fe-B magnets can be made using the same procedures as Sm-Co magnets. Indeed, the Sumitomo Special Metals initial alloy was created using the induction melting method. However, because of the challenges connected with the commercial manufacturing of NdFe-B using R-D/KOR technologies, the General Motors Corporation created an alternate production process. This process relies on the melt spinning technique to rapidly solidify the molten alloy. The melt solidifies as flakes with microscopic crystallites (less than 10 nm in diameter), which are crushed into platelet-shaped particles and annealed to form a powder with extremely high coercivity. By aligning the particles with a magnetic field, the resulting isotropic tiny polycrystalline powder particles cannot be used to make anisotropic magnets. However, isotropic bonded magnets may be made commercially by mixing the magnetic powder with a binder (such as polymer or rubber), molding or pressing, and then curing. Although the binder dilutes the magnetic qualities greatly, most RE-based bonded magnets outperform most non-earth magnets; this is due to the RE magnets' high magnetic properties. Hot pressing can also be used to create densification without the use of magnets. Furthermore, advances in hot deformation techniques enabled the production of well-oriented magnets with energy products as high as 40 MGOe (318 kJ/m3), which are essentially similar to the best-sintered magnets. These magnets are more corrosion resistant than sintered magnets and may have greater long-term performance [24,32–34].

3.5   Hexaferrite/ferrite based magnets

The Philips group found hexaferrite with qualities appropriate for permanent magnets in the 1950s. These ceramic oxides are employed in a broad variety of electronic and electrical applications due to their high resistivity and low eddy current losses. They are also utilized for shielding equipment from electromagnetic noise signals, as well as radar absorbers in military applications, and they have ideal characteristics as microwave absorbers. When compared to SmCo and NdFeB permanent magnets, ferrite permanent magnets have lower magnetic characteristics and energy products. The following magnetic properties characterise commercially available high performance ferrite magnets: $B_r = 0.41$ T, $H_{cJ}=$ 335 kA/m, $H_{cB} = 300$ kA/m, and $(BH)_{max} = 35$ kJ/m^3. Throughout the twentieth century, strong research efforts led in the creation of high-performance modern permanent magnets and a noticeable increase in the maximum energy product. Additionally, research efforts increased permanent magnets' other magnetic characteristics, which are crucial for a broad variety of practical applications. The magnetic property ranges of the top commercially available permanent magnets that currently dominate the world market. Despite having lower magnetic characteristics than RE-based magnets, ferrite magnets have advantages over Re-Co and NdFeB magnets in terms of inexpensive production costs, raw material availability, and high corrosion resistance at high working temperatures (> 250°C). The shortage of RE-elements, as well as the monopoly imposed on the world market by Chinese producers after 1985, as well as China's strategy of limiting RE-element export owing to internal demand, all contributed to an exponential rise in RE market prices to nearly prohibitive levels. While ferrite sintered magnets may be purchased for $5 to $20 per kilogram in US and European markets, sintered RE-based magnets can be purchased for up to $200 per kilogram. The market price of RE magnets makes them less cost-effective than ferrite magnets at a flux density of about three times higher. In addition, increasing the surface area of the weaker ferrite magnet can provide flux densities comparable to RE magnets in an air gap. However, this may result in a greater machine volume, which can be addressed through creative machine design. Furthermore, ferrite magnets can be employed in powder form for real-world applications, reducing machining and processing costs. Because of the advantages of ferrite magnets versus RE-based magnets, ferrite magnets have been the market's top PMs since 1987. According to a 2014 report, ferrite magnets accounted for more than 80% of the PM market volume in 2012. Despite their superior performance, the high costs of raw materials, manufacturing, and processing SmCo and NdFeB magnets appear to be prohibitive. However, since ferrite magnets are less expensive, compared to RE-based magnets, ferrites accounted for just over 20% of global PM market sales, whereas Alnico and Sm-Co magnets (with only 3% of global market volume) accounted for around 8% of global PM market revenues. Nd-Fe-B magnets dominated the global PM market in 2012, and the 62 percent market share forecast in 2010 has expanded dramatically. Despite the high cost of Nd-Fe-B magnets, their market share has been steadily increasing since their introduction. This is because device shrinking necessitates high-performance magnets with high flux and energy products. The number of patents and the volume of published scientific research on permanent magnet materials may reflect the level of interest among scientists and developers, as well as the product's economic viability. This graph shows an exponential increase in published scientific work, which corresponds to the growing demand for PMs. In comparison to the other magnetic materials, Nd-Fe-B magnets have dominated scientific study during the last three decades, suggesting strong scientific interest in producing and describing Nd-Fe-B magnetic materials. Since the discovery of ceramic oxide in the early 1950s, the number of published hexaferrite articles has grown exponentially. Since the discovery of these magnets, the number of Nd-Fe-B PM patents has increased dramatically, indicating the rapid development of these materials over the last three decades. Although Nd-Fe-B magnets were discovered fifty years after alnico magnets, they have a similar number of patients presently and are anticipated to surpass alnico's overall number of patents shortly [33].

4. Tasks for improving the hard ferrite-based magnets

Rare-earth elements are critical raw resources that are subject to supply constraints, price volatility, and environmental concerns. As a consequence, significant effort is being made to lessen our reliance on them, which directly impacts the PM value chain. Several initiatives are being pursued, including increasing the recyclability of rare-earth-based materials and devices, producing competitive rare-earth-free electronics, and finding alternative magnets to replace NdFeB and SmCo-based magnets. Ferrites are an attractive possibility for replacement since they are the most extensively utilised magnet on a volumetric basis. Ferrites' magnetic characteristics, on the other hand, must be enhanced in order for them to be used in a broader variety of applications. Naturally, this doping is often feasible in applications requiring magnets with low $(BH)_{max}$ (50–200 kJ/m^3), but ferrite doping may be useful in certain situations. Wind turbine and vehicle generators come into this category, and the development of competitive technologies based on the phenomenon of enhanced ferrites might result in widespread doping. As a consequence of this requirement, the condition of ferrite magnets is improving. In sintered ceramics, hexaferrites have a $(BH)_{max}$ value of around 40 kJ/m^3. The bulk of the H_C values used to establish $(BH)_{max}$ have been frozen for more than 40 years. Given both barium and strontium M-type hexaferrites have much larger magnetic anisotropy fields, 1.7 T and 1.8 T, respectively, than H_C and B_S, both sintered and bonded commercial magnets have coercive fields comparable to remanence and approaching saturation induction (0.4 T). In this case, $(BH)_{max}$ is close to the highest possible value, $BS_2/0$. As a consequence, the first strategy for enhancing the $(BH)_{max}$ is to improve the saturation induction. However, for permanent magnets, HC should be greater than Nef/Ms to avoid demagnetization, where Nef is the demagnetizing factor caused by the form of the magnet. As a consequence, a secondary objective is to raise the coercive field such that HC >Nef/Ms is maintained. Because the H_C is proportional to the anisotropy field and hence to $K_1/\mu_0 M_s$, enhancing both the H_C and M_S with a single-phase material requires an additional increase in the magnetic anisotropy, K_1. On the other hand, altering the magnetic structure might have unintended consequences for the ceramics' characteristics. The first is a modification of the exchange contacts, which results in a modification of the Curie temperature, T_C. When the Curie temperature is lowered, the magnetization may decay more rapidly than in conventional ceramics. This may jeopardise the use of magnets in high-temperature environments. Another consideration is that the negative thermal coefficient of hexaferrite causes the coercive field to increase with temperature. As a consequence, ferrite magnets often perform better at temperatures more than 100°C than they do at room temperature. The temperature dependence of HC is impacted by the relative temperature dependences of K_1 and M_S [14], which should vary as a result of T_C changes during doping. As a consequence, any changes to any component may have an adverse effect on the ultimate usable temperature range.

5.   Parameters responsible for improving the performance of the hard ferrites for their utilization in permanent magnets application

There are number of factors responsible for the improvement of hard ferrites performance as explained below.

5.1   Influence of size at the nanoscale

The reduction of grain size to the nanoscale is generally recognised as one of the most efficient methods of increasing a magnetic material's coercivity. Alnico alloys are a kind of nanocomposite material that is one of the most effective types of permanent magnets [14]. Alnico alloys are formed of elongated ferromagnetic nanograins scattered across a weakly magnetic Al–Ni matrix, and they continue to play an important role in the permanent magnet industry 90 years after their discovery. Using an external magnetic field during the annealing process may enhance the energy product of the material [35]. Magnetization reversal in bulk Ferro/Ferri-magnetic materials may be caused by a number of events, including domain wall pinning and reverse domain nucleation and propagation. Additionally, as the material's volume is compressed, a threshold size is reached below which the formation of domain walls is no longer energetically advantageous and a single domain spin configuration is produced. The single-domain particle's magnetic activity is represented by a large magnetic moment $= M_S V$, where V is the volume. As a consequence, the individual atomic spins undergo a coordinated rotation, reversing their orientation. The competition between three energy variables, magnetic anisotropy constant, magnetostatic energy, and exchange energy, defines the critical diameter, dC, for a given material's single domain spin configuration state.

The d_C is given as for a spherical nanocrystal [14]:

$$d_C = (18E_\sigma) / (\mu_0 M_s^2) = (36\sqrt{AK}) / (\mu_0 M_s^2)$$

Where μ_0 denotes vacuum permeability and $E = 2(AK)^{1/2}$ denotes a domain's surface energy. For the majority of magnetic materials, average d_C values vary from a few tens to many hundreds of nm. Experimental and theoretical estimations for Ba-ferrite imply that d_C is around 500 nm, however larger values up to 1 μm have been reported. González et al. determined the activation volume, or the size of the nucleus responsible for magnetization reversal, of a series of Ba-ferrite particles with diameters ranging from 65 nanometers to micrometres [36]. The coercive field rose from 200 kA/m to 320 kA/m as the particle size decreased below 800 nm, whereas the activation size decreased from 40 nm to 25 nm throughout the series. The scientists discovered that in particles bigger than 800 nm, domain wall propagation is responsible for reversal, but coherent rotation is responsible for reversal in smaller particles. The transition from multidomain to single-domain is not simple, and several non-uniform reversal modes, such as curling and buckling, may occur. Where, despite improbable logical reversal, the exchange energy does not stay constant. For example, Chan et al. used magnetic force microscopy to measure the switching field of a series of isolated nanoparticles of doped barium ferrite and discovered that the Stoner–Wohlfarth model was valid only for average sizes up to 50 nm; once this size was exceeded, incoherent reversal modes caused the switching field to decrease [37]. Additionally, using electron holography, examination of the magnetic domain structure on isolated Ba-ferrite particles obtained by heating a combination of Fe_2O_3, BaO, and Bi_2O_3 to 1100°C showed the coexistence of single-domain and two-domain magnetic states across the full examined range of 0.1–2 m. As a consequence of the size dependence of the reversal process, coercivity increases as grain size decreases, as schematically shown in Figure 4, as long as the single domain threshold is met [14]. As the nanoparticles' size reduces, they become increasingly sensitive to demagnetization by heat, until magnetic irreversibility is lost and the superparamagnetic state is obtained. The size of the grain above which superparamagnetism may be seen for a given measurement duration is governed by its structure and crystal quality (the presence of defects and impurities). It is often assumed that SPM relaxation becomes dominant for Ba and Sr-ferrite at diameters smaller than 40-60 nm. Additionally, bear in mind that these constraints apply to the time scales used in laboratory experiments (about 102–104 s), which are substantially less than the shelf life required for a permanent magnet (the reduction of magnetization must be lower than 5 percent over 100 years).

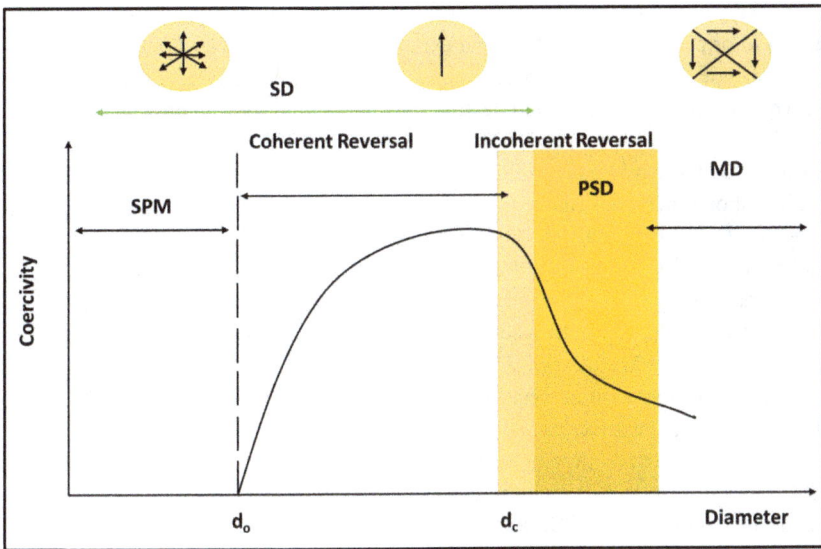

Figure 4. The room temperature change of coercivity with grain size from superparamagnetic state to single domain state and multidomain states is depicted in the diagram. PSD represents the pseudo-single domain area, and the light grey zone denotes the size range where incoherent reversal processes are feasible [14].

A comprehensive examination of a series of polycrystalline $BaFe_{12}O_{19}$ samples synthesised using a typical solid-state reaction confirmed the anticipated trend of coercivity increasing with particle size. The particle size was varied between 500 nm and several microns by increasing the calcination temperature from 1000 to 1400 °C [38]. According to the authors, coercivity is regulated by pinning of the magnetization at grain boundaries, which rises with the reciprocal of the size. In comparison, they asserted that nucleation-induced coercivity and domain wall pinning are negligible in larger multi-domain grains. On the other hand, our study demonstrated how grain size may significantly affect the coercivity of Ba-ferrite, while also confirming that the critical single domain size is about 500 nm.

5.2   Influence of the shape (Morphology)

Along with particle size, the form of the particles plays a significant influence in determining the coercive field. For noninteracting single domain spheroidal particles, the anisotropy field, HA, may be defined as $H_A = H_K + D_N M_S$, where $H_K = 2K_1/\mu_0 M_S$ is the magnetocrystalline anisotropy and D_N is the demagnetizing term [39]. If the easy axis is parallel to or perpendicular to the main axis of the spheroid, it has a positive or negative value. M-type ferrites have a regular hexagonal thin platelet structure with the c-axis aligned with the magneto-crystalline easy axis along the short dimension. This shape is related to the production of 001 crystallographic faces with a low energy. Additionally, depending on the synthesis conditions, acicular platelets are commonly formed. When the magnetization is perpendicular to the easy axis, a substantial demagnetizing field emerges. $H_A = H_K - D_N M_S$ is the magnetic anisotropy decrease due to the form effect. Although D_N is a tensor, it may be treated as a scalar in this example for convenience. The diameter (D) to thickness (t) ratio determines the D_N term: $D_N = 1$ when t << D, as in thin platelets; it then decreases with D/t, until reaching zero for equiaxial morphology (D/t = 1). When D/t<1 is increased, D_N shifts and H_A increases in relation to H_K. This talk emphasises the critical role of proper particle shape design in maximising the ferrite's effective anisotropy. Numerous writers have conducted experiments to demonstrate the shape's significance. Kubo et al., for example, investigated the effect of particle shape on the coercivity of Ba–ferrite nanoparticles synthesised through the glass-crystallization process, both pure and doped with trace amounts of Co and Ti [40]. They observed that by increasing the diameter (d) to thickness (t) ratio, HC decreased to approximately 32 kA/m for d/t values of around 6.

On the contrary, the platelet form makes it simpler to orient small particles (less than 100 nm) physically or magnetically. This is critical because grain orientation along the c-axis allows for a rise in decreased remanence from half of saturation magnetization. Which is the value discovered for isotropic magnets manufactured from powders containing single domain, randomly oriented particles (0.5 is the theoretical value proposed by the Stoner-Wohlfart model and corresponds to (BH)$_{max}$ values of around 8 kJ m^{-3}) up to values close to 1 [39]. Additionally, the Stoner–Wohlfart model shows that the coercive field increases with orientation: while $H_C = 0.49\ H_A$ for randomly oriented particles, H_C equals H_A for directed particles. Due to the platelet structure, particles may be oriented easily, and coercive field and remanence can be enhanced, facilitating the transition from powders to bulk magnets. Platelets may also be stacked to generate a similar effect. Interparticle interactions mandate a reduction in the internal field of platelets in this situation, which may enhance their magnetic properties.

5.3 Fabrication techniques for the preparation of hard ferrite-based nanomaterials

The capacity to control the shape and size of particles down to the nanoscale is thus important for producing high-performance permanent magnets. As a consequence, efforts to synthesis M-hexaferrite nanoparticles, nanofibers, and nanorods have been documented in the literature dating all the way back to the 1980s, at the dawn of the nanomaterial era. Additionally, it should be emphasised that the bulk of current research has focused on the fabrication of nanoscale M-hexaferrites with properties suitable for applications other than permanent magnets. M-type hexagonal ferrites have a wide range of applications in technology, including microwave devices, magnetic recording, and electromagnetic wave absorbers. Various techniques have been developed to date, including solid-state synthesis, milling, sol-gel, co-precipitation, aerosol pyrolysis, microemulsion, combustion synthesis, glass crystallisation, and hydrothermal processes [14]. Low-temperature chemical techniques, in particular, were shown to be effective in the formation of M-hexaferrite nanostructures with controlled size and shape and excellent magnetic properties. Sol-gel methods have been extensively employed to synthesise unsubstituted and substituted Sr- and Ba-ferrite nanoparticles. Particular emphasis has been placed on simplifying the function of each manufacturing parameter. As a result, it has been shown that careful management of the different parameters is required to minimise the formation of secondary phases such as hematite or goethite, to control the size, and to optimise the magnetic properties. Nga et al. studied the influence of pH, the Fe/Sr molar ratio, the calcination temperature, and the length of calcination on the structure and magnetic properties of $SrFe_{12}O_{19}$ particles synthesised using a sol-gel process using metal nitrates in citric acid as precursors [41]. They observed that after calcination at 850°C, pure nanometric hexagonal particles with a Fe/Sr molar ratio of 10.5 and a pH of 1 could be formed. The particles had MS = 56 A m^2/kg and HC = 525 kA/m. Using the sol-gel auto combustion technique, it was established that a Ba/Fe molar ratio of 11.5 is optimal for producing a well-crystalline M-type Ba–ferrite powder with a crystallite size of 22 nm and enhanced magnetic properties (H_C = 420 kA/m). Strict control of the manufacturing parameter is also required when employing the hydrothermal approach to fabricate nanoscale M-type hexaferrites. This manufacturing approach has mostly been shown to be successful in accomplishing this goal. Additionally, the hydrothermal method, like the sol-gel technique, is simple and cost-effective, and it can easily scaled up to industrial manufacture. Despite these advantages, multiple investigations have shown that the final materials' features are controlled by a range of synthesis parameters, many of which are interdependent, such as temperature, precursor molar ratio, duration, concentration, and base nature.

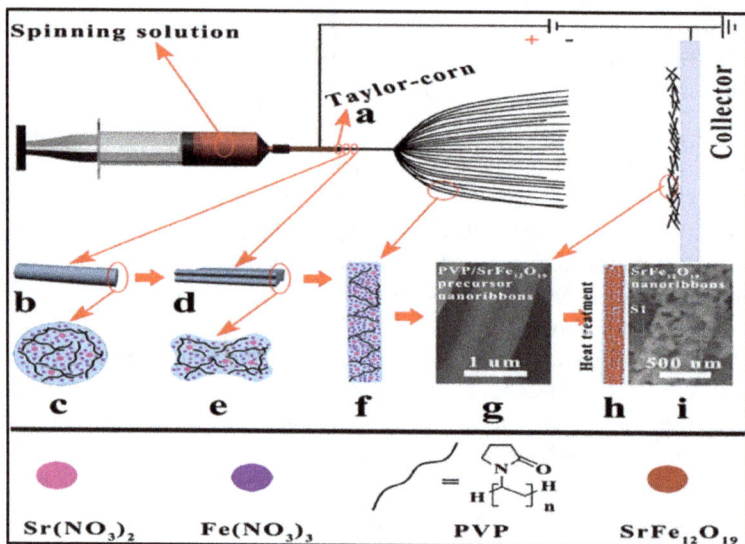

Figure 5. The fabrication mechanism of SrFe$_{12}$O$_{19}$nanoribbons is depicted schematically [43].

On the other hand, although this feature increases the technique's ability to create hexaferrites with unique properties for each application, it also complicates establishing the optimal process parameters to get the required result. In this regard, Saura-Muzquiz et al. recently reported some interesting results revealing the vital importance of the metal ion solution concentration when using NaOH as a precipitating agent in the precursor preparation for the hydrothermal formation of Sr–ferrite nanoparticles [42]. By increasing the iron concentration from 0.05 to 0.75 M, hexagonal platelet-like particles with mean diameters ranging from 48 to 254 nm were obtained with magnetic properties that were highly dependent on average size (specific remanence, R, ranging from 20 to 31 Am^2/kg and HC ranging from 68 to 175 kA/m). They were able to manufacture SrFe$_{12}$O$_{19}$ magnets with a high density (more than 95% of the bulk density) using spark plasma sintering.

5.3.1 One dimensional nanostructure

1D magnetic nanostructures have also garnered considerable attention, since their large surface area and high aspect ratio may significantly increase the total magnetic anisotropy, therefore radically altering the coercivity. Numerous attempts have been undertaken in this field to fabricate mono-dimensional hexagonal M-hexaferrite structures, with electrospinning emerging as one of the most promising techniques known to far. $SrFe_{12}O_9$ ribbons with an average width of 484±15 to 1099±18 nm were successfully created utilising a polyvinylpyrrolidone sol-gel based electrospinning approach followed by an air heat treatment, as depicted in Figure 5 [43].

$BaFe_{12}O_{19}$ nanofibers of average length 1.5 m were also synthesised using electrospinning a homogeneous polyvinyl pyrrolidone/barium nitrate/iron nitrate precursor sol-gel solution, followed by calcination at 800°C. These nanofibers are composed of randomly distributed, single nanoparticles with an average diameter of 70 nm stacked along the nanofiber axis. The researchers investigated the development process and proposed a model consisting of a series of phase transitions occurring at increasing temperatures, with Fe_2O_3, BaO, $BaCO_3$, and $BaFe_2O_4$ serving as intermediate species [44].

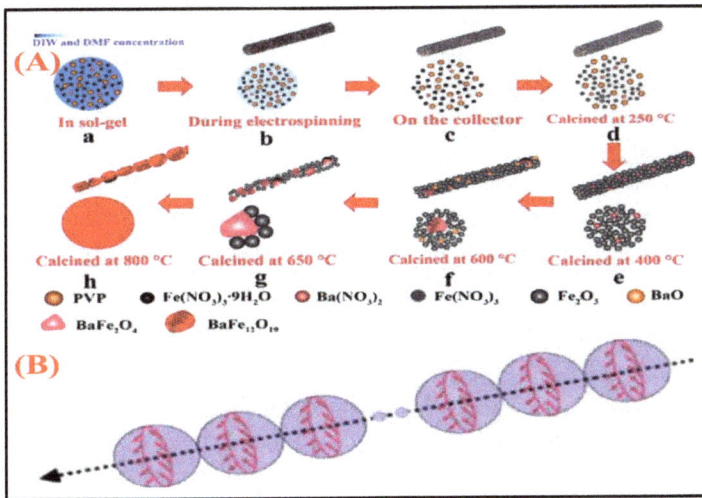

Figure 6. (A) The production mechanism of $BaFe_{12}O_{19}$ single-particle-chain nanofibers is depicted schematically. The concentrations of deionized water and N,N-dimethylformamide drop from blue to white. (B) The magnetization reversal mechanism is represented schematically [44].

At room temperature, the resulting nanofibers had an S value of 71.5 Am^2/kg, a lowered remanence of roughly 0.5, and an H_C value of 473 kA/m, all of which were less than the theoretically predicted values. The magnetization reversal mechanism of $BaFe_{12}O_{19}$ single-particle-chain nanofibers was examined theoretically and found to be consistent with a curling model.

5.4   Controlling substitution in the structure of hard ferrites

The M-hexaferrite structure enables simple enhancement of their magnetic properties by substituting for some of the cations in particular positions of the magnetoplumbite structure. Kojima evaluated the field's development up until the early 1980s [45]. Furthermore, until the recent overviews by Lisjak et al., [46], Pullar [10], and Mahmood et al., [20], this topic lacked reviews. We will focus on those studies that demonstrate higher magnetization, coercive field, or anisotropy field values than the best commercial hexaferrite magnets. The intrinsic magnetic structure of this magnetoplumbite crystal is determined by the presence of five sites and their magnetic coupling through oxygen-driven super-exchange interactions, and doping M-type ferrites is often used to alter this intrinsic magnetic structure. The spin and orbital moments of cations are influenced by the site and charge dispersion, which is dictated by the spin configuration (low, high, or intermediate spin), which also impacts the ion's local anisotropy and net magnetic moment. This configuration may be used to meet a variety of replacement scenarios. uOne option is to alter total magnetization. Another way to improve magnetocrystalline anisotropy is to increase local magnetic anisotropy owing to the orbital moment's contribution or changes in chemical and crystal structures. Generally, all of the ions contained in the structure have been investigated for possible replacement. Nevertheless, the most popular technique includes cation substitution, while altering the oxygen anions has also been attempted. The major conclusion drawn from the literature is that the changes in characteristics caused by doping are exceedingly difficult to anticipate and comprehend [47].

5.4.1   Enhancing magnetization (M_s) by substitution

Three strategies have been used to increase the magnetization of M-hexaferrites by doping: replacement with a single magnetic element, substitution with a single nonmagnetic element, and substitution with several magnetic elements (non-magnetic and magnetic elements). Using ab initio calculations, Dixit et al. predicted that Bi, Sb, Sn, and Sc ions might enhance magnetism [48]. Regrettably, no experimental result has yet resulted in an increase in magnetism. Al^{3+} has been widely investigated, but to no result. Wang et., al. study of magnetic Ni-doped $BaFe_{12}O_{19}$ synthesised through sol-gel is the only work that suggests an enhancement for a single magnetic cation [49]. They found a specific magnetization of 101 Am^2/kg for x = 0.8. The authors argue that the origin is due to the presence of Ni^{2+} in $4f_2$ spin down sites, as well as Fe^{2+} in $4f_1$ and $4f_2$ spin down sites.

5.4.2 Doing substitution in hexaferrites with large anisotropy and coercive field

Considering the majority of works that sought to enhance magnetization lowered anisotropy, we now analyse works that improved the latter. A significant number of studies have reported enhancements in the H_C of M-hexaferrites by evaluating the influence of various cations. The focus of this study will be on chosen research that describes materials with coercive fields greater than those of the finest commercial $SrFe_{12}O_{19}$ and $BaFe_{12}O_{19}$ magnets, i.e., greater than 400 kA/m (0.5 T) [50–52]. Dixit et. al., suggest from theoretical calculations that doping with P, Co, Al, Ga, and Ti cations should enhance the anisotropy constant or field [48]. The Al-doped ferrites are especially fascinating since they have higher H_C and induction magnetizations than NdFeB ($H_C \sim$ 1.19 MA/m, 1.5 T, and $\mu_0 M_S$ = 1.6 T) in some situations. As a result, these materials will not be demagnetized by the stray field produced by the NdFeB magnets and may be used in magnetic systems that contain RE magnets. So, summaries, the ferrite doping approach provides for advancements in the characteristics of M-type hexaferrites, such as increased magnetization or coercive field. Because the cationic occupy the spin-down sites, Ni and Zn-based doped ferrites have higher magnetizations. In comparison, because the cations occupancy spins up sites, Al-based ferrites have a high H_C owing to their low magnetization. Commercial LaCo doped ferrites are an intermediate case in which Co cations occupy both spin up and down sites, resulting in a weak change in magnetization, but the La-induced contraction of the crystal structure and the orbital moment of the Co^{2+} cation cause an increase in anisotropy and thus in the coercive field. Furthermore, the ultimate hysteresis of the magnets is determined by the submicrometric size of the powders and the platelet shape. The fact that LaCo ferrites, which display a little improvement in B_r and H_c over undoped ferrites, have made their way into commercial applications highlights the importance of minor increases in hexaferrite characteristics.

Considering that the bulk of efforts to increase magnetization resulted in decreased anisotropy, we now examine efforts to increase the latter. Numerous research have shown improvements in the H_C of M-hexaferrites by examining the effect of different cations. This study will concentrate on selected research that describes materials with coercive fields larger than those of the best commercial $SrFe_{12}O_{19}$ and $BaFe_{12}O_{19}$ magnets, i.e., more than 400 kA/m (0.5 T) [50–52]. Dixit et al. believe that doping with P, Co, Al, Ga, and Ti cations should result in an increase in the anisotropy constant or field [48]. Al-doped ferrites are particularly intriguing since they exhibit stronger H_C and induction magnetizations than NdFeB in specific conditions (HC $\sim$1.19 MA/m, 1.5 T, and $\mu_0 M_S$ =

1.6 T). As a consequence, these materials will not get demagnetized by the stray field generated by the NdFeB magnets and may be employed in magnetic systems including RE magnets. Thus, the ferrite doping strategy enables improvements in the properties of M-type hexaferrites, such as higher magnetism or coercive field. Ni and Zn-based doped ferrites exhibit greater magnetizations due to the cationic occupying the spin-down sites. In contrast, Al-based ferrites have a high H_C because to their low magnetism, which is caused by the cations occupied spinning up sites. Commercial LaCo doped ferrites are an intermediate case in which Co cations occupy both spin up and down sites, resulting in a weak change in magnetization, but the contraction of the crystal structure and orbital moment of the Co^{2+} cation results in an increase in anisotropy and thus in the coercive field. Additionally, the magnets' final hysteresis is dictated by the sub-micrometric size of the powders and the form of the platelets. The fact that LaCo ferrites, which exhibit a little increase in Brand H_c over undoped ferrites, have found their way into commercial applications demonstrates the critical significance of small improvements in hexaferrite properties.

Concluding remarks

In the current chapter, we have presented a comprehensive latest research work about the hard ferrites for permanent magnets application. The different synthesis approaches have been employed for making suitable hard/hexaferrites for PMs application respectively. The various parameters responsible for improving the performance of hard ferrites such as effect of size at the nano level, morphologically effect, effect of substitution either divalent or trivalent within the crystal structure were taken into practice. Alternatively, in other way, the alternative approach of substituting cations at the specified interstitial sites of magnetoplumbite crystal structure helps us in the modification of characteristics of hard ferrites for their utilization in permanent magnets application.

References

[1] J.S. Smart, Effective field theories of magnetism, Saunders, 1966. https://doi.org/10.1063/1.3048415

[2] S. Elliott, The physics and chemistry of solids, Wiley, 1998.

[3] W. Nolting, A. Ramakanth, Quantum theory of magnetism, Springer Science & Business Media, 2009. https://doi.org/10.1007/978-3-540-85416-6

[4] A. Aharoni, Introduction to the Theory of Ferromagnetism, Clarendon Press, 2000.

[5] P. Weiss, Hypothesis of the molecular field and ferromagnetic properties, J. Phys. 6 (1907) 661-690. https://doi.org/10.1051/jphystap:019070060066100

[6] K.H.J. Buschow, F.R. Boer, Physics of magnetism and magnetic materials, Springer, 2003. https://doi.org/10.1007/b100503

[7] B.D. Cullity, C.D. Graham, Introduction to magnetic materials, John Wiley & Sons, 2011.

[8] J.B. Carlson, Lodestone compass: Chinese or olmec primacy?: Multidisciplinary analysis of an olmec hematite artifact from san lorenzo, veracruz, mexico, Science. 189 (1975) 753-760. https://doi.org/10.1126/science.189.4205.753

[9] K.H.J. Buschow, New permanent magnet materials, Materials Science Reports. 1 (1986) 1-63. https://doi.org/10.1016/0920-2307(86)90003-4

[10] R.C. Pullar, Hexagonal ferrites: a review of the synthesis, properties and applications of hexaferrite ceramics, Progress in Materials Science. 57 (2012) 1191-1334. https://doi.org/10.1016/j.pmatsci.2012.04.001

[11] V.G. Harris, A. Geiler, Y. Chen, S.D. Yoon, M. Wu, A. Yang, Z. Chen, P. He, P.V. Parimi, X. Zuo, Recent advances in processing and applications of microwave ferrites, Journal of Magnetism and Magnetic Materials. 321 (2009) 2035-2047. https://doi.org/10.1016/j.jmmm.2009.01.004

[12] Ü. Özgür, Y. Alivov, H. Morkoç, Microwave ferrites, part 1: fundamental properties, Journal of Materials Science: Materials in Electronics. 20 (2009) 789-834. https://doi.org/10.1007/s10854-009-9923-2

[13] Ü. Özgür, Y. Alivov, H. Morkoç, Microwave ferrites, part 2: passive components and electrical tuning, Journal of Materials Science: Materials in Electronics. 20 (2009) 911-952. https://doi.org/10.1007/s10854-009-9924-1

[14] C. de Julian Fernandez, C. Sangregorio, J. de la Figuera, B. Belec, D. Makovec, A. Quesada, Topical review: progress and prospects of hard hexaferrites for permanent magnet applications, Journal of Physics D: Applied Physics. (2020). https://doi.org/10.1088/1361-6463/abd272

[15] X. Obradors, X. Solans, A. Collomb, D. Samaras, J. Rodriguez, M. Pernet, M. Font-Altaba, Crystal structure of strontium hexaferrite SrFe12O19, Journal of Solid State Chemistry. 72 (1988) 218-224. https://doi.org/10.1016/0022-4596(88)90025-4

[16] D. Holtstam, U. Hålenius, Nomenclature of the magnetoplumbite group, Mineralogical Magazine. 84 (2020) 376-380. https://doi.org/10.1180/mgm.2020.20

[17] D. Makovec, B. Belec, T. Goršak, D. Lisjak, M. Komelj, G. Dražić, S. Gyergyek, Discrete evolution of the crystal structure during the growth of Ba-hexaferrite nanoplatelets, Nanoscale. 10 (2018) 14480-14491. https://doi.org/10.1039/C8NR03815E

[18] G.D. Soria, P. Jenus, J.F. Marco, A. Mandziak, M. Sanchez-Arenillas, F. Moutinho, J.E. Prieto, P. Prieto, J. Cerdá, C. Tejera-Centeno, Strontium hexaferrite platelets: a comprehensive soft X-ray absorption and Mössbauer spectroscopy study, Scientific Reports. 9 (2019) 1-13. https://doi.org/10.1038/s41598-019-48010-w

[19] K. Momma, F. Izumi, VESTA: a three-dimensional visualization system for electronic and structural analysis, Journal of Applied Crystallography. 41 (2008) 653-658. https://doi.org/10.1107/S0021889808012016

[20] S.H. Mahmood, I. Abu-Aljarayesh, Hexaferrite permanent magnetic materials, in: Materials Research Forum LLC, 2016. https://doi.org/10.21741/9781945291074

[21] J.J. Becker, Rare-Earth-Compound Permanent Magnets, Journal of Applied Physics. 41 (1970) 1055-1064. https://doi.org/10.1063/1.1658811

[22] I.A. Al-Omari, R. Skomski, R.A. Thomas, D. Leslie-Pelecky, D.J. Sellmyer, High-temperature magnetic properties of mechanically alloyed SmCo 5 and YCo 5 magnets, IEEE Transactions on Magnetics. 37 (2001) 2534-2536. https://doi.org/10.1109/20.951226

[23] S.H. Mahmood, High performance permanent magnets, in: Hexaferrite Permanent Magnetic Materials, Materials Research Forum LLC, Millersville, PA, 2016: pp. 47-73. https://doi.org/10.21741/9781945291074

[24] K.J. Strnat, Modern permanent magnets for applications in electro-technology, Proceedings of the IEEE. 78 (1990) 923-946. https://doi.org/10.1109/5.56908

[25] K. Kamino, Y. Kimura, T. Suzuki, Y. Itayama, Variation of the Magnetic Properties of Sm (Co, Cu) 5 Alloys with Temperature, Transactions of the Japan Institute of Metals. 14 (1973) 135-139. https://doi.org/10.2320/matertrans1960.14.135

[26] J.J. Croat, J.F. Herbst, Melt-spun R0. 4Fe0. 6 alloys: Dependence of coercivity on quench rate, Journal of Applied Physics. 53 (1982) 2404-2406. https://doi.org/10.1063/1.330826

[27] N. Koon, B. Das, J. Geohegan, Composition dependence of the coercive force and microstructure of crystallized amorphous (Fe x B 1-x) 0.9 Tb 0.05 La 0.05 alloys, IEEE Transactions on Magnetics. 18 (1982) 1448-1450. https://doi.org/10.1109/TMAG.1982.1061968

[28] N.C. Koon, B.N. Das, Magnetic properties of amorphous and crystallized (Fe0. 82B0. 18) 0.9 Tb0. 05La0. 05, Applied Physics Letters. 39 (1981) 840-842. https://doi.org/10.1063/1.92578

[29] G.C. Hadjipanayis, R.C. Hazelton, K.R. Lawless, New iron-rare-earth based permanent magnet materials, Applied Physics Letters. 43 (1983) 797-799. https://doi.org/10.1063/1.94459

[30] M. Sagawa, S. Fujimura, N. Togawa, H. Yamamoto, Y. Matsuura, New material for permanent magnets on a base of Nd and Fe, Journal of Applied Physics. 55 (1984) 2083-2087. https://doi.org/10.1063/1.333572

[31] J.J. Croat, Observation of large room-temperature coercivity in melt-spun Nd0. 4Fe0. 6, Applied Physics Letters. 39 (1981) 357-358. https://doi.org/10.1063/1.92728

[32] M.A. Sweeney, F.C. Perry, J.R. Asay, M.M. Widner, Shock effects in particle beam fusion targets, in: AIP Conference Proceedings, American Institute of Physics, 1982: pp. 188-192. https://doi.org/10.1063/1.33357

[33] M.J. Kramer, R.W. McCallum, I.A. Anderson, S. Constantinides, Prospects for non-rare earth permanent magnets for traction motors and generators, Jom. 64 (2012) 752-763. https://doi.org/10.1007/s11837-012-0351-z

[34] O. Gutfleisch, M.A. Willard, E. Brück, C.H. Chen, S.G. Sankar, J.P. Liu, Magnetic materials and devices for the 21st century: stronger, lighter, and more energy efficient, Advanced Materials. 23 (2011) 821-842. https://doi.org/10.1002/adma.201002180

[35] L.H. Lewis, F. Jiménez-Villacorta, Perspectives on permanent magnetic materials for energy conversion and power generation, Metallurgical and Materials Transactions A. 44 (2013) 2-20. https://doi.org/10.1007/s11661-012-1278-2

[36] J.M. González, C. De Julian, A.K. Giri, S. Castro, M. Gayoso, J. Rivas, Magnetic viscosity and microstructure: Particle size dependence of the activation volume, Journal of Applied Physics. 79 (1996) 5955-5957. https://doi.org/10.1063/1.362118

[37] T. Chang, J.-G. Zhu, J.H. Judy, Method for investigating the reversal properties of isolated barium ferrite fine particles utilizing magnetic force microscopy (mfm), Journal of Applied Physics. 73 (1993) 6716-6718. https://doi.org/10.1063/1.352512

[38] J. Dho, E.K. Lee, J.Y. Park, N.H. Hur, Effects of the grain boundary on the coercivity of barium ferrite $BaFe12O19$, Journal of Magnetism and Magnetic Materials. 285 (2005) 164-168. https://doi.org/10.1016/j.jmmm.2004.07.033

[39] E.C. Stoner, E.P. Wohlfarth, A mechanism of magnetic hysteresis in heterogeneous alloys, Philosophical Transactions of the Royal Society of London. Series A,

Mathematical and Physical Sciences. 240 (1948) 599-642.
https://doi.org/10.1098/rsta.1948.0007

[40] O. Kubo, T. Ido, H. Yokoyama, Y. Koike, Particle size effects on magnetic properties of BaFe12- 2 x Ti x Co x O19 fine particles, Journal of Applied Physics. 57 (1985) 4280-4282. https://doi.org/10.1063/1.334585

[41] T.T.V. Nga, N.P. Duong, T.T. Loan, T.D. Hien, Key step in the synthesis of ultrafine strontium ferrite powders (SrFe12O19) by sol-gel method, Journal of Alloys and Compounds. 610 (2014) 630-634. https://doi.org/10.1016/j.jallcom.2014.04.193

[42] M. Saura-Múzquiz, C. Granados-Miralles, H.L. Andersen, M. Stingaciu, M. Avdeev, M. Christensen, Nanoengineered high-performance hexaferrite magnets by morphology-induced alignment of tailored nanoplatelets, ACS Applied Nano Materials. 1 (2018) 6938-6949. https://doi.org/10.1021/acsanm.8b01748

[43] P. Jing, J. Du, J. Wang, L. Pan, J. Li, Q. Liu, Width-controlled M-type hexagonal strontium ferrite (SrFe 12 O 19) nanoribbons with high saturation magnetization and superior coercivity synthesized by electrospinning, Scientific Reports. 5 (2015) 1-10. https://doi.org/10.9734/JSRR/2015/14076

[44] J. Zhang, J. Fu, F. Li, E. Xie, D. Xue, N.J. Mellors, Y. Peng, BaFe12O19 single-particle-chain nanofibers: preparation, characterization, formation principle, and magnetization reversal mechanism, Acs Nano. 6 (2012) 2273-2280. https://doi.org/10.1021/nn204342m

[45] H. Kojima, Fundamental properties of hexagonal ferrites with magnetoplumbite structure, Handbook of Ferromagnetic Materials. 3 (1982) 305-391. https://doi.org/10.1016/S1574-9304(05)80091-4

[46] D. Lisjak, A. Mertelj, Anisotropic magnetic nanoparticles: A review of their properties, syntheses and potential applications, Progress in Materials Science. 95 (2018) 286-328. https://doi.org/10.1016/j.pmatsci.2018.03.003

[47] G.F. Dionne, Magnetic oxides, Springer, 2009. https://doi.org/10.1007/978-1-4419-0054-8

[48] V. Dixit, S.-G. Kim, J. Park, Y.-K. Hong, Effect of ionic substitutions on the magnetic properties of strontium hexaferrite: A first principles study, AIP Advances. 7 (2017) 115209. https://doi.org/10.1063/1.4995309

[49] M. Wang, Q. Xu, J. Liu, Z. Wang, N. Ma, P. Du, Extra up-spin magnetic moments and extraordinary high saturation magnetization of Ni2+ doped barium ferrite in 4f2

site, Materials Research Express. 6 (2019) 086104. https://doi.org/10.1088/2053-1591/ab1a0c

[50] G. Albanese, A. Deriu, Magnetic properties of Al, Ga, Sc, In substituted barium ferrites: a comparative analysis, Ceramurgia International. 5 (1979) 3-10. https://doi.org/10.1016/0390-5519(79)90002-4

[51] M. Awawdeh, I. Bsoul, S.H. Mahmood, Magnetic properties and Mössbauer spectroscopy on Ga, Al, and Cr substituted hexaferrites, Journal of Alloys and Compounds. 585 (2014) 465-473. https://doi.org/10.1016/j.jallcom.2013.09.174

[52] T.B. Ghzaiel, W. Dhaoui, A. Pasko, F. Mazaleyrat, Effect of non-magnetic and magnetic trivalent ion substitutions on BaM-ferrite properties synthesized by hydrothermal method, Journal of Alloys and Compounds. 671 (2016) 245-253. https://doi.org/10.1016/j.jallcom.2016.02.071

An Introduction to Hard Ferrites: From Fundamentals to Practical Applications   Materials Research Forum LLC
Materials Research Foundations **142** (2023) 152-184   https://doi.org/10.21741/9781644902318-6

Chapter 6

Hard Ferrites for High Frequency Antenna Applications

Asha Kumari, Rahul Sharma*

Career Point University, Vill: Tikker-Kharwarian, Tehsil Bhoranj, Distt. Hamirpur, Himachal Pradesh, India 176041

*rsharma8886@gmail.com

Abstract

Advances in wireless communication place an increasing number of demands on antenna performance, necessitating the presence of various capabilities in a single device. Reconfigurable antennas are frequently utilized to meet these various application demands within a restricted area. The purpose of this book chapter is to summarize general introduction about hard ferrites, different synthesis methods of ferrite for antenna application and altering functioning of antenna by reconfiguring them by ferrites. Along with this we have also focused on miniaturization and reconfiguration of antennas which is becoming a very important aspect of wireless communication devices. Miniaturization and reconfiguration of antennas involves a deliberate alteration in the form and/or electrical behaviour of the antenna, leading in a change in the antenna's functioning.

Keywords

Miniaturization, Hard Ferrites, Reconfiguration, Electrical Behaviour, Hard Ferrites

Contents

1.   Introduction

Antennas are anticipated to have varied performances and features depending on the application. One of the most essential characteristics to examine is efficiency, and efficiency is directly determined by losses. There are several elements that impact the performance of various types of antennas operating in various frequency bands. The mismatch loss and antenna size, for example, affect performance in the (High frequency) HF band (3 MHz-30 MHz). The dielectric loss of the substrate material is one of the most critical variables in the microwave frequency range, aside from mismatch loss. In the millimetre wave frequency spectrum, however, when all of the sub-antennas are perfectly matched, the loss in the feeding network is the component that limits antenna gain and efficiency.

Different strategies and methods are studied to increase the performances, taking into account diverse influencing variables. To begin, small and effective ferrites are chosen to increase the HF antennas' performance. To increase the performance of the HF antenna, effective ways of utilising ferrites are presented. The ferrites may be placed effectively and efficiently in locations with high current density by analysing the current in the ground plane of the antenna, reducing the effect of image current. Furthermore, by using ferrite loadings, the antenna's bandwidth is considerably enhanced while the antenna's size is reduced.

The frequently utilized high-loss material is the major factor influencing the antenna's efficiency for microstrip antennas that function in the microwave band. It's a commercial technique for putting planar periodic structures in the ground plane or on the patch. The antenna's used power is lowered, and the antenna's bandwidth is enhanced, thanks to periodic features in the ground plane. The periodic structures can operate in the same way as the patch when the planar structures are positioned regularly on it. From the slots in the patch, more energy may be emitted. The resonant frequencies can be found by examining the dispersion relationship. The antenna's bandwidth is substantially improved by using periodic features in the patch.

The loss in the millimetre wave antenna's feeding network results in a small number of antenna components in an array, limiting the array's gain. Instead of using a standard microstrip line, a low-loss transmission technique called substrate integrated waveguide is proposed for the feed network. Furthermore, the antenna array's shape is simplified. The antenna array's losses have been considerably reduced. As a result, the antenna array's gain is enhanced. Varied antenna optimization approaches are necessary due to the different frequency bands. The modelling findings are compared to the measurement data for the high-performance antennas. The antennas have great performance, efficiency, and are compact in size.

1.1 Ferrites for antenna application

Different types of antennas operating in various frequency bands have better performance. There have been a number of noteworthy contributions. Using commercial ferrite tiles substantially improves the performance of HF antennas. The mutual resistance of the antenna is studied using the image current theory. The ferrite is utilised successfully to decrease the influence of the picture current by evaluating the strength and dispersion of the image current. Commercial ferrite is also utilised to lower the antenna's height. The lower bound of bandwidth is increased by loading the ferrites, and HF antenna downsizing is achieved. The microwave antenna's performance is increased by etching periodic features.

The efficiency and bandwidth of the patch antenna are enhanced while the antenna size is reduced by etching the features in the ground plane. More energy may be emitted into free space and the bandwidth is improved by employing periodic structures as the patch. The size of the HF antenna is enormous due to the large wavelength of the HF band. As a result, it is preferable to minimise the HF antenna's size. However, by reducing the antenna's size, the antenna's efficiency is also diminished. The platform is positioned in relation to the lower bound of the HF band's wavelength. The effective size of the antenna is increased by activating partial or entire size of the platform, resulting in enhanced effectiveness at lower frequencies. However, the platform's geometry and size have a significant impact on the design's performance. This approach is not applicable to all antennas and platforms. The antenna is based on a folded dipole that is electrically tiny. The antenna's efficiency is improved by connecting two dipoles with an 180° phase shifter. The balanced dipole antenna, like other horizontal polarised antennas, must be set high enough to function. To decrease mismatch loss and enhance efficiency at lower frequencies, commercial ferrites are utilised to cover the whole ground plane. The huge number of commercial ferrites, on the other hand, raises the total weight and price. The antenna's major source of energy is magnetic. By utilising ferrites to lower the amount of energy held in the antenna, more energy may be radiated out, increasing the antenna's overall efficiency.

The discovery of stones that would attract iron began the history of ferrite materials in different years before the advent of Christ. The minerals magnetite were called after the region of Magnesia in Asia Minor, where vast quantities of these stones were discovered (Fe_3O_4). The first usage of magnetite was in the form of "Loadstones," which were used by early navigators to determine magnetic north [1]. William Gilbert published the first scientific study on De magnete magnetism around 1600 [2]. Hans Christian Oersted proved in 1819 that an electric current in a wire influences the needle of a magnetic compass, a discovery that was later refined by other scientists and led to further discoveries in the subject of electromagnetism. Ferrite is derived from the Latin word ferrum, which means iron. Ferrites are now a homogenous ceramic substance made up of different oxides with iron oxide as the primary ingredient. Yogoro Kato and Takeshi Takei of Tokyo Institute of Technology published the first ferrite compound in 1930 [3].

Soft and hard ferrites are divided into two groups depending on their magnetic coercivity and resistance to demagnetization [4]. Soft ferrite, such as zinc, cobalt, nickel, manganese, and magnesium ferrites, do not keep their magnetic after being magnetised, but hard ferrites, also known as permanent magnets, may maintain their magnetism after being magnetised. However, in the majority of ferrites study, scientists categorise ferrites based on their crystal structure. Ferrites are divided into four categories: Spinal ferrites, garnet ferrites, hexa ferrites, and orthoferrites. Spinal ferrites, also known as $FeO.Fe_2O_3$, are a kind of naturally occurring ferrite. Mineral spinal ($MgAl_2O_4$ or $MgO.Al_2O_3$) crystallises in the cubic system, resulting in the spinal structure [5]. Braggs [6] and Nishikawa [7] were the first to discover this crystal structure. $MeO.Fe_2O_3$ or $MeFe_2O_4$ is the general formula, where Me is a divalent metal ion.

Garnet ferrites are ferrites that can hold a big trivalent rare earth with a lot of magnetic moments. The structure of the silicate mineral garnet is shown through garnet ferrites. Similar to natural garnet, magnetic garnet crystallises in a dodecahedral or 12-sided shape. $Me_3Fe_5O_{12}$ is the general formula. Hexagonal ferrites class of magnetic oxide suggest magnetoplumbite structure which comes from the mineral of the same name. Hexagonal ferrites are defined by $MeFe_{12}O_{19}$ where Me is usually Ba,Sr or Pb Orthoferrites are also known as Perovskites. The formula if $RFeO_3$ where R is yttrium or rare earth ion [8]. These are also cubic ferrites with slightly distorted structure. Fig.1 describing classification of ferrites based on crystal structure and magnetic properties.

FERRITES CLASSIFIACTION

Classification based on Crystal Structure	Classification Based on Magnetic Field
➤Spinal Ferrites ➤Hexagonal Ferrites ➤Garnet Ferrites ➤Ortho ferrites	➤Soft Ferrites ➤Hard Ferrites

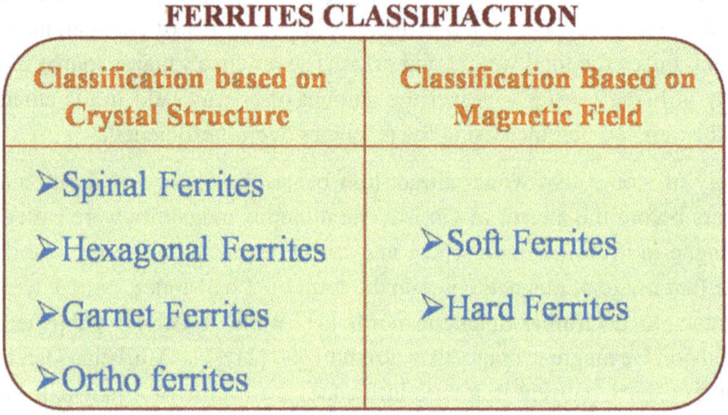

Figure 1. Classification of ferrites based on crystal structure and magnetic properties.

With the fast growth of mobile communication networks, numerous services for mobile phone apps have become necessary and, as a result, realised. Antennas have been miniaturised as a result of miniaturisation and high-efficiency trends. With merely a circuit application, these antenna gains were restricted.

A variety of techniques have been used to miniaturise antennas, including meander lines or slots, dielectric loading, and so on [9]. The antenna's size is reduced, resulting in a reduction in bandwidth. The use of meander lines or slots to reduce antenna size reduces antenna performance such as bandwidth and efficiency. The dielectric loading approach reduces antenna performance and hence restricts permittivity growth. To reduce antenna gain, a thick dielectric substrate either boosts surface wave energy or stores it within itself. Magneto-dielectric materials provide hope for overcoming the basic constraints of electrically tiny antennas. Relative permittivity (ε) and relative permeability (μ) are connected to the electrical wavelength, size, and bandwidth of an antenna [10]. An antenna's operational frequency and bandwidth are determined as follows by Eq. 1 and Eq. 2:

$$f = \frac{c}{2L\sqrt{\epsilon\mu}} \tag{1}$$

$$BW = \frac{96\frac{t}{\lambda_0}\sqrt{\frac{\mu}{\varepsilon}}}{\sqrt{2}[4+17\sqrt{\mu\varepsilon}]} \tag{2}$$

Where, c is the velocity of light, L is the length of a patch, t is the thickness of the substrate and $\lambda 0$ is the wavelength in free space.

High permittivity or high permeability materials would be desirable for antenna size reduction based on the operating frequency and bandwidth equation, but they would also lower the bandwidth or raise the loss tangent, respectively. Materials with relatively high and comparable permittivity and permeability values, on the other hand, can shrink in size without reducing bandwidth. Because they may flip between functionalities inside a single structure without having to use numerous antennas, reconfigurable antennas have become a popular technique. Antenna reconfiguration entails an intentional modification in the antenna's shape and/or electrical behaviour, resulting in a change in the antenna's functionality [11]. Typically, such antennas have two or more states that may be switched discretely or continuously. These various states are usually achieved by modifying the antenna's current pathways, either by rearranging the antenna or by altering the antenna's surrounding medium. Many current radiofrequency (RF) systems, such as wireless and satellite communication, imaging, and sensing, employ reconfigurable antennas [12].

The issue has gotten a lot of interest since Schaubert [13] filed the first patent on reconfigurable antennas in 1983. Various designs have been presented in the literature to accomplish reconfigurability in terms of frequency, radiation pattern, polarisation, or a combination of two or three of the previously mentioned. Various reconfiguration strategies that may be integrated into an antenna design in order to redistribute its surface current through changes in the feeding network, the physical structure of the antenna, or the radiating edges can be used to achieve the necessary reconfigurability [14]. As illustrated in Fig. 2, reconfiguration approaches may be classified into various types. Electrical, optical, physical, and material reconfigurations are the four primary types of reconfiguration procedures employed. To redirect surface currents, electrically reconfigurable antennas employ RF microelectromechanical systems (MEMS), PIN diodes, or varactors. Optically reconfigurable antennas are those that use photoconductive switching components. Mechanical deformation can be used to change the antenna shape in the physical reconfiguration approach.

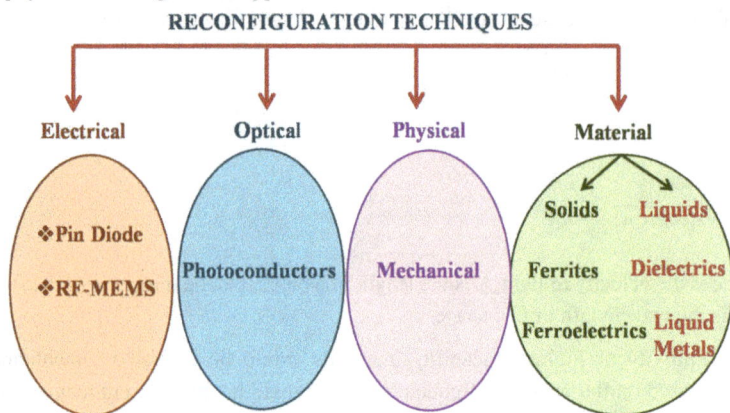

Figure 2. Different strategies used to reconfigure Antennas.

Finally, the reconfiguration of an antenna may be accomplished using smart materials like as ferrites, ferroelectrics, liquid metal, liquid crystal, and liquid dielectrics [15]. Hexagonal Ba-ferrites are frequently recommended as tiny antenna materials. M-type ($BaFe_{12}O_{19}$), W-type ($BaM_2Fe_{16}O_{27}$, M: Co, Zn, etc.), Y-type ($Ba_2M_2Fe_{12}O_{22}$), and Z-type ($Ba_3M_2Fe_{24}O_{41}$) are the different types of Ba-ferrites [16]. The Co_2Z ferrite ($Ba_3Co_2Fe_{24}O_{41}$) has a relatively high permeability and resonance frequency among these Ba ferrites. However, because to its complicated crystal structure, the single phase of Co_2Z ferrite is difficult to manufacture [17]. Kim et al. [18] investigated the structural and microwave characteristics of Mn added Co_2Z ferrite, $Ba_3Co_{2-2x}Mn_{2x}Fe_{24}O_{41}$ (0.1 x 0.5. The solid-state reaction technique was used to make all of the magnetic ceramics for small antenna applications. Their studies revealed that size of an antenna can be reduces by magneto dielectric materials.

2. Synthesis of hard ferries for antenna applications

2.1 Various synthesis methods

Co-precipitation, salt-melt, ion exchange, sol-gel, citrate, hydrothermal, combustion, microemulsions, and other processes are used to make ferrites [19-23]. Other combination synthesis approaches, such as the sol gel-citrate method, the sol gel-hydrothermal technique, and others, exist in addition to these synthesis methods. Processing procedures have a major impact on the physical characteristics of ferrites, which also have a substantial impact on magnetic properties. Below are some of the chemical approaches that are addressed.

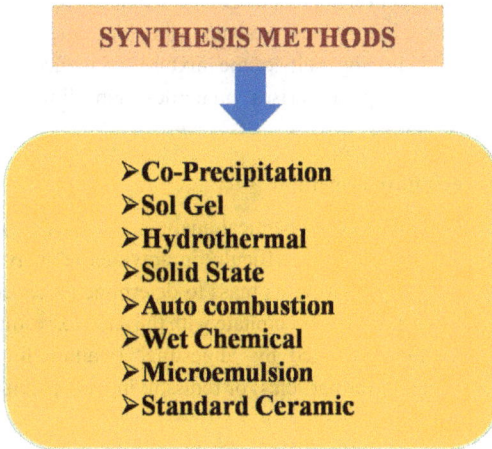

Figure 3. Different methods for synthesis of hard ferrites for antenna applications.

2.1.1 Ceramic powder milling method

The ferrites are synthesised by heating a combination of barium carbonate powders and oxides. After that, the ceramic powder is processed to obtain the finer material. After a lot of milling, we can get nanosize. This approach does not allow for the creation of single-phase ferrites. During the processing, impurities are picked up and material is lost. Due to the poor reactivity of the starting components, processing requires a high temperature and a lengthy time [24]. Which allows nanoparticles to engage in thermal motion and gives the potential of their self-assembly into superstructures by finally reaching the thermodynamic optimum [25] while converting ferrites into ultrafine nano agglomerates.

2.1.2 Reaction in solid state method

This approach entails combining hydroxide, oxide, carbonate, or sulphate raw materials that have been heated to around 1100 °C for an extended period of time, resulting in homogeneous microstructures, aberrant grain development, poor sintering behaviour, and uncontrolled cation stoichiometry [26]. A combination of barium carbonate and iron oxide is burned at extremely high temperatures (1150–1250 °C) in this process. The powder is then pulverised, which can result in impurities in the powder as well as stains in the crystal lattices, affecting the magnetic characteristics [27, 28]. Chemical decomposition processes, in which a mixture of solid reactants is heated to form a new solid composition and gases, are used in the solid-state reaction approach. Complex oxides are often made from simple oxides, carbonates, nitrates, hydroxides, oxalates, alkoxides, and other metal salts using this process. To enhance the homogeneity of the mixture and reduce the particle size of the powder, the technique usually comprises numerous annealing phases with several intermediate milling operations.

2.1.3 Chemical coprecipitation method

Coprecipitation is a crucial subject in chemical analysis, where it may be both unwanted and advantageous. Coprecipitation is a difficulty in gravimetric analysis, which involves precipitating the analyte and measuring its mass to determine its concentration or purity. Unwanted contaminants frequently coprecipitate with the analyte, resulting in excess mass. This difficulty is frequently alleviated by "digesting" (waiting for the precipitate to equilibrate and create larger, purer particles) or redissolving and precipitating the material again.

Salts are chemically co-precipitated using a base (NaOH). Since the early 1960s, this approach has been employed [28, 29]. The mixture is magnetically stirred and heated to around 70°C. The particles are then centrifuged and rinsed with deionized water. The granules are finally dried.

Inclusion, occlusion, and adsorption are the three primary processes of coprecipitation. An inclusion (incorporation in the crystal lattice) happens when an impurity occupies a lattice site in the carrier's crystal structure, resulting in a crystallographic defect; this can happen when the impurity's ionic radius and charge are comparable to the carrier's. An adsorbate is an impurity that is weakly or firmly attached to the precipitate's surface (adsorbed). When an adsorbed impurity becomes physically stuck inside the crystal as it develops, it is called an occlusion.

2.1.4 Sol gel synthesis method

A homogenous solution of metal salts is prepared and put into a solution of citric acid in sol-gel synthesis. Ethylene glycol, an organic coordinating agent, is added to the solution to make a sol that forms a gel structure after the water evaporates. Ammonia solution is used to maintain pH. At the nanoscale, this approach is employed to create more complicated hexaferrites [30]. The sol-gel process has a variety of benefits, including a low annealing temperature and superior microstructure control [31]. Chemical decomposition processes, which require heating a combination of solid reactants to form a new solid composition and gases, are used in the solid-state reaction method. Simple oxides, carbonates, nitrates, hydroxides, oxalates, alkoxides, and other metal salts are typically utilised in this process to produce complicated oxides. To enhance the homogeneity of the mixture and reduce the particle size of the powder, the technique typically comprises numerous annealing phases with multiple intermediate milling operations.

Extra milling makes the powder more active in the heat-treatment procedures that follow (i.e., more sinter active). The solid-state method is less costly and requires less complicated equipment. Furthermore, vast quantities of powder may be made in a very straightforward method. However, as compared to the wet powder production methods mentioned below, the resultant powder has a high degree of agglomeration, resulting in large particle size and restricted homogeneity.

2.1.5 Temperature specific combustion synthesis

Solution combustion synthesis (SCS) has become one of the most used processes for producing a wide range of oxide materials. The SCS method offers several advantages, including speed, ease of use, control over many elements of the products, production of high purity products, stabilisation of metastable phases, and the ability to create products of practically any size and shape. Environmental cleanup with SCS-derived powders is particularly successful. It has been established that a single-step combustion process may deposit nanocrystalline ionic catalysts on ceramic cordierite honeycombs.

On a heated plate, a solution of salts, ammonia, and citric acid is evaporated to a dry powder, with the pH adjusted to 7. Citric acid polymerizes, releasing carbon dioxide and entirely converting cations to iron oxide and barium carbonate. Because of the tiny grain size, the material formed has weak magnetic characteristics [32]. Aqueous combustion synthesis (ACS) or low temperature combustion synthesis (LCS) can be used to carry out this procedure [33]. After that, the mixture is auto-combusted in a microwave oven, yielding nano powder [34].

2.1.6 Hydrothermal synthesis method

One of the most frequent ways for preparing nanomaterials is hydrothermal synthesis. It's essentially a solution-reaction-based method. The creation of nanomaterials in hydrothermal synthesis may take place over a wide range of temperatures, from ambient temperature to extremely high temperatures. Depending on the vapour pressure of the primary component in the reaction, either low-pressure or high-pressure conditions can be utilised to regulate the morphology of the materials to be synthesised. This method has been used to successfully synthesis a wide range of nanomaterials.

Hydrothermal synthesis has a number of benefits over other methods. At high temperatures, hydrothermal synthesis can produce nanomaterials that are unstable. The hydrothermal process can generate nanomaterials with high vapour pressures with minimal material loss. In hydrothermal synthesis, the compositions of nanomaterials to be generated may be well controlled by liquid phase or multiphase chemical processes. This special issue provides as a venue for presenting the most recent research findings in hydrothermal nanomaterial production.

In hydrothermal synthesis a solution of metal salts and base are made under pressure to give the product. The unreactive precursors are washed away with dilute HCl. Ataie et al. studied the effects of many bases such as NaOH, KOH and NH4OH in the synthesis of hexaferrites [35].

2.1.7 Wet chemical method

It is feasible to generate specific surface structures, phases, forms, and sizes of metal oxide nanoparticles through wet chemical synthesis, resulting in a set of desired attributes. To get the required nanomaterials, wet chemical synthesis pathways allow careful control of the reaction conditions (temperature, substrate concentration, additives or surfactants, pH, etc.). Chemical reactions in the solution phase are dealt with employing precursors under controlled experimental settings in wet-chemical synthesis techniques. Each wet-chemical synthesis process is unique, hence there is no such thing as a universal norm for these types of synthesis methods. These synthesis procedures have been utilised to make 2D nanomaterials that are difficult to make using top-down methods. For 2D nanomaterial manufacturing, wet-chemical synthesis techniques offer a high degree of controllability and repeatability.

For the manufacture of hexaferrites, the wet chemical process is more effective than the ceramic method. The raw components are mixed together in a homogenous manner in this process. Processing requires a low temperature and sintering takes a lengthy time. During the procedure, no material is lost. The good news is that there's a better chance of single-phase development [24].

2.1.8 Microemulsions method

Particles that can be controlled a precipitation in water-in-oil microemulsion technique was used to make starch nanoparticles of various sizes. Microemulsions have ultralow interfacial tension, a high interfacial area, and are thermodynamically stable, resulting in monodispersed nanoparticles. Because of their special features, including as ultralow interfacial tension, huge interfacial area, thermodynamic stability, and the ability to dissolve otherwise immiscible liquids, microemulsions have a wide range of functions and applications in the chemical and biological sciences. Nanoparticles have essential technical uses, such as catalysts, high-performance ceramic materials, microelectronic devices, high-density magnetic recording, and medication delivery, in addition to being of basic scientific interest.

Mixing two immiscible liquids and stabilising them with an artificial coating of surface-active substances yields nanoparticles. This approach [36] allows for morphological control. Separation, washing, and drying of the precipitates follow.

3.   Different compositions of hard ferrites for antenna applications

Due to the huge size of the antenna, miniaturisation in electronics has virtually reached a limit, and decrease in the physical size of radio frequency (RF) systems is still a grey area [37]. By raising frequency, the size of the antenna may be decreased, but the cost of electronics rises, making it an unfeasible option in many sectors needing mass manufacturing. Small antennas are becoming increasingly important in communication, medical, space, and defence applications. Size and bandwidth are the primary constraints in missiles, satellites, military, and other space-based applications [38]. Microstrip antennas have grown in popularity in recent years due to its planar structure, conformal design, tiny size, low manufacturing cost, and simple fabrication procedure [39]. Table 1 elaborate the previously reported compositions of hard ferrites along with other parameter for antenna applications. Dielectric materials are used as the substrate in traditional microstrip antennas. Because of the low permittivity of the substrate materials, antennas must be big in order to fit into the given area. The use of a high-permittivity substrate can reduce the size of the device, but it reduces the gain, radiation pattern, and bandwidth. This is due to the field's confinement in a high-permittivity area, as well as an impedance mismatch with the surrounding medium [40].

Table 1: Previously reported compositions of hard ferrites along with other parameter for antenna applications.

Sr. No.	Composition	Permittivity(ε)	Permeability(μ)	Size (nm)	Bandwidth	Ref.
1	Cobalt ferrite LDPE	~1–2.905	~1.01–1.05	10 nm	9.5 GHz	44 (2012)
2	$Ba_3Co_{2-2x}Mn_{2x}Fe_{24}O_{41}$	19.774	15.183	--	510 MHz	18 (2014)
3	$Ni_{1-x}Zn_xFe_2O4$ (NiZn), $Ni_{0.8x}Zn_xCo_{0.2}Fe_{1.98}O_{4-\delta}$ (NiZnCo)	--	~0.9–2.7	--	200 MHz–6 GHz	46 (2015)
4	$Ni_{0.7}Zn_{0.3}Fe_2O_4$ + x $BaFe_{12}O_{19}$ (x=30)	15	~15.5	--	75-570 MHz	45 (2014)
5	$(Mn_{0.5}Zn_{0.35}Co_{0.15}Fe_2O_4$ + $SrFe_{12}O_{19})$	4.1–0.1	3.72–0.28	--	1 GHz	47 (2016)
6	$BaFe_{12}O_{19}$	6.2-0.04	1.9 -0.18	70 nm	68-166 MHz	48 (2017)
7	$(Mn_xCo_{1-x}$ Fe_2O_4/PAA, 0-x-1)	--	--	6-12 nm	75 MHz	50 (2018)
8	$BaFe_{12}O_{19}$-$CoFe_2O_4$	15	13	--	500 MHz	49 (2020)
9	$SrFe_{12}O_{19}$–Li_2MoO_4	5-6.7	1.10-1.16	--	12.89 GHz	52 (2021)
10	$Mg_{0.7}Cd_{0.3}Fe_{1.92}Ga_{0.08}O_4$	21-30	15-19	0.42-1.18μm	1-200 MHz	51 (2021)
11	$Co_xZn_{(0.90-x)}Al_{0-10}Fe_2O_4$	3.5-4.5	0.90-1.0	20-26 nm	2.5 to 5.5 GHz (S-band)	53 (2022)

Using magneto-dielectric materials, these restrictions can be addressed [41]. Magneto-dielectric materials are non-natural materials that exhibit both dielectric and magnetic characteristics. Microwave specialists have been intrigued by these materials for two reasons: (i) They give two degrees of freedom in altering material characteristics, namely relative permittivity (ε_r) and relative permeability (μ_r), and therefore the flexibility to design the material that is best suited for a particular application. (ii) They have a high dielectric strength at high frequencies, which results in low eddy current losses, making them ideal for high-frequency applications. Ferrites are one example of this type of material. Proper raw material selection and synthesis methods can affect their structural and electromagnetic characteristics [42]. As a result, a moderate value of μ_r and ε_r can be used to miniaturise the antenna [43].

In 2012 Borah and Bhattacharyya carried out research to determine the complicated permittivity and permeability of different volume fractions of magnetodielectric composites containing cobalt ferrite nano inclusions, as well as magnetic characterisation [44]. In deionized water, cobalt (II) nitrate hexahydrate (98 percent pure, $Co (NO_3)_2.6H_2O$) and iron (III) nitrate nonahydrate (P98 percent pure, $Fe (NO_3)_2.9H_2O$) were dissolved in stochiometric proportions. Surfactant oleic acid ($C_{17}H_{33}COOH$) and 3M NaOH are added and mixed up to precipitation after steady stirring for 2.5 hours at 80°C. The resulting polycrystalline precipitate is annealed for three hours at 600 °C. Cobalt ferrite nanoparticles had an average grain size of ~10 nanometers. With an increase in inclusion content from 1% VF to 5 percent VF, the actual part of permittivity and permeability of the samples changes from ~1–2.905 to ~1.01–1.05. The imaginary portion of permeability and the tanδ of permittivity are determined to be on the order of ~10^{-3} and ~10^{-1}, respectively.

Dielectric characteristics were evaluated using a contact superstrate method. The complicated permeability of the materials is studied using the cavity perturbation technique. Vibrating sample magnetometry is used to determine the 4nMs value and coercivity. The size and homogenous distribution of nano inclusions are determined by examining the structural and surface morphologies of composite materials. Cobalt ferrite nanoparticles had an average grain size of ~10 nanometers. With an increase in inclusion content from 1% VF to 5% VF, the actual portion of permittivity and permeability of the samples changed from ~1–2.905 to ~1.01–1.05. The omnidirectional behaviour of the antenna was revealed by the directivity values determined from the radiation characteristics. Low side lobe will assist MIMO (Multiple-In Multiple-Out) applications reduce interference in radiation patterns. The MPA's (microstrip rectangular patch antenna) radiation performance may be improved even further by using external magnetic bias.

For a tiny antenna application, the sintering behaviour and magneto-electric characteristics of $Ba_3Co_{2-x}Mn_{2x}Fe_{24}O_{41}$ ($0.1 \leq x \leq 0.5$) ceramics were studied by Kim et al. 2014 [18]. All $Ba_3Co_{2-x}Mn_{2x}Fe_{24}O_{41}$ ceramics were sintered at 1250°C using the solid state reaction technique. Most sintered specimens, on the other hand, produced a single Z-type phase. With increasing frequency, the real portion of permittivity dropped and the loss tangent of permittivity grew. There was no discernible variation in component ratio. Over 500 MHz, the actual component of permeability dropped significantly, with an inflection point about 800 MHz; it also dropped with Mn additions. The $Ba_3Co_{2-x}Mn_{2x}Fe_{24}O_{41}$ ceramics actual portion of permeability was greater than that of the $Ba_3Co_2Fe_{24}O_{41}$ ceramics. The permeability loss tangent increased with frequency but did not change significantly with composition ratio. At 510 MHz, the $Ba_3Co_{0.2}Mn_{0.8}Fe_{24}O_{41}$ ceramics sintered at 1250°C had real part permittivity, loss tangent of permittivity, and real part permeability, loss tangent of permeability of 19.774, 0.176 and 15.183, 0.073, respectively. The operating frequency of the simulated PIFA using $Ba_3Co_{0.2}Mn_{0.8}Fe_{24}O_{41}$ ceramics was 541 MHz, while the impedance bandwidth (Voltage Standing Wave Ratio, VSWR) was 39 MHz (7.2 percent). In the operation band, the total efficiency of the suggested PIFA was around 25%. Based on these findings, it appears that the size of an antenna may be decreased without sacrificing bandwidth by utilising magneto-dielectric materials instead of FR4 (designation FR-4 applies to glass-reinforced epoxy laminate materials).

Zheng et al. 2013 [45] used solid-state reaction technique to successfully manufacture a variety of NiZn ferrite composites with different $BaFe_{12}O_{19}$ hexaferrite (BaM) additions (x) for their prospective application as magneto-dielectric antenna substrate materials. When a combination of NiZn ferrite and BaM was sintered at 1200°C, a W-type hexagonal phase was produced, and a diphase composite ferrite consisting of NiZn spinel ferrite and BaM hexaferrite was eventually created, according to XRD and energy-dispersive spectrum studies. BaM considerably reduces the frequency dispersion of the permittivity spectrum, resulting in a constant permittivity ε' of about 15 for doped samples from 1MHz to 1 GHz, which is confirmed to be closely connected to refined grains according to Koops' hypothesis. Furthermore, the magnetic and dielectric losses in the doped samples were shown to be lower than in the Undoped NiZn ferrite. In the sample with x = 30% wt percent, nearly identical values of μ and ε were obtained. Authors suggested that a magneto-dielectric composite with decreased physical dimensions and strong impedance matching to free space might be a potential choice for antenna construction. This research might also lead to a novel approach for fabricating magneto-dielectric materials with the requisite electromagnetic properties.

Mattei et al. 2015 [46] studied the magnetocrystalline anisotropy constants (K1) and saturation magnetostriction constants (S) of $Ni_{1-x}Zn_xFe_2O^4$ (NiZn) and $Ni_{0.8-x}Zn_xCo_{0.2}Fe_{1.98}O_{4-\delta}$ (NiZnCo) ferrites suitable for antenna downsizing. The observed results for NiZn ferrites were consistent with published data, indicating that both the experimental procedure and the suggested modelling of stress-induced FMR (Ferrimagnetic Resonance) changes were valid. The ability of $Ni_{0.6}Zn_{0.2}Co_{0.2}Fe_{1.98}O_4$ (which had the greatest natural ferrimagnetic resonance and the highest λ_S among the ferrites tested) to be utilised as a substrate for antenna miniaturisation at frequencies up to 1 GHz has been proven.

Saini et al. 2016 [47] proposed the miniaturisation of a microstrip patch antenna utilising a composite nanosized ferrite material. The effect of increasing the relative permeability of the substrate material on the physical size and efficiency of a microstrip antenna was studied using detailed simulations. On the basis of the comprehensive simulation, an analytical formula for estimating the effective relative permeability was constructed. The electromagnetic characteristics of a composite nano ferrite ($Mn_{0.5}Zn_{0.35}Co_{0.15}Fe_2O_4$ + $SrFe_{12}O_{19}$) with an average crystallite size of 72 nm were investigated. Co-precipitation was used to prepare the substrate material. The electromagnetic characterisation yielded matching values of complex permittivity ($\varepsilon^* = 4.1–0.1j$) and complex permeability ($\mu^* = 3.72–0.28j$) up to 1 GHz. In comparison to a pure dielectric FR4 substrate, simulation and test findings show that an antenna constructed using the following specifications may lower the patch size by almost 44 % and enhance the reflection loss bandwidth by -10 dB. Authors suggested that composite nano ferrites of $Mn_{0.5}Zn_{0.35}Co_{0.15}Fe_2O_4 + SrFe_{12}O_{19}$ are strong contender for a high-bandwidth miniaturised antenna in the microwave frequency range.

In the design of proximity fuzes, calculating the exact height of burst has always been a difficulty. Saini et al. 2017 [48] proposed miniaturisation of the patch antenna using barium hexaferrite ($BaFe_{12}O_{19}$) as the substrate material. Radio frequency-based sensors can be designed for this purpose, but the size and bandwidth of the antenna increases the design complexity; thus, this study proposed miniaturisation of the patch antenna using barium hexaferrite ($BaFe_{12}O_{19}$) as the substrate material. The structural and electromagnetic characteristics of the nanohexaferrite substrate material were determined using a wet chemical technique. X-ray diffraction revealed a crystallite size of 60 nm on average. The antenna construction built and simulated reveals that the size of the antenna may be decreased by up to 42.5% using the electromagnetic characteristics of synthesised magneto-dielectric material. It also boosts the bandwidth of the antenna on the (Flame retardant) FR4 substrate from 68 to 166 MHz. As a result, $BaFe_{12}O_{19}$ is offered as a potential choice for a downsized, high-bandwidth antenna for proximity fuzes.

For high-frequency antenna applications, Polley et al. created a $BaFe_{12}O_{19}$-$CoFe_2O_4$ ferrite composite [49]. Nanocomposites were synthesized by precipitation method. The existence of both BaM and CoF ferrite phases in composite ferrite was revealed by XRD analysis. The composites were sintered for 4 hours at 1100°C/4 h. In comparison to the other composites, the composite with $x = 0.25$ composition had the most compact microstructure structure and greater bulk density. Due to greater densification, the composite (with x = 0.25) also had the best permittivity (~15), permeability (~13), and antenna miniaturisation factor (~14) at 500 MHz. The permeability of the composite was nearly constant up to 500 MHz. The composite with x = 0.25 composition had the lowest coercivity, showing that the harsh magnetic behaviour of BaM ferrite changed to a soft magnetic character during composite production, possibly due to the development of larger BaM grains in the composite.

Alcala et al. 2017 [50] investigated the electrical response of toroidal coils containing mixed ferrites magnetic nanoparticles (MNPs) implanted in a polyacrylamide matrix ($Mn_xCo_{1-x}Fe_2O_4$/PAA, $0 \le x \le 1$). The MNPs were made by thermal breakdown of molecular precursors, and the $Mn_xCo_{1-x}Fe_2O_4$/PAA toroidal cores were made by copolymerization of MNPs with acrylamide and bis-acrylamide. The cubic spinel phase was represented by

An Introduction to Hard Ferrites: From Fundamentals to Practical Applications   Materials Research Forum LLC
Materials Research Foundations **142** (2023) 152-184   https://doi.org/10.21741/9781644902318-6

MNP X-Ray Diffraction (XRD) patterns. The average size of MNPs measured by Transmission Electron Microscopy (TEM) and was in between 6 and 12 nm. To compare the results, authors measured the properties of a commercial toroidal coil and discovered that the impedance curves for each configuration (commercial and Laboratory-made coils) showed a resonance peak around 75 MHz; the signal intensity of the Laboratory-made coil increases by one order of magnitude over the commercial coil. Both magnetic and electrical measures were shown to be linked to manganese content. The benefit of the proposed $Mn_xCo_{1-x}Fe_2O_4$/PAA toroidal coils system is that the flexible combinations of Mn^{2+} and Co^{2+} components allowed for easy tweaking of the electrical and magnetic characteristics in order to optimise the coils impedance.

Gan et al. 2021 [51] studied the impact of sintering temperature (900°C, 915°C, 930°C, 945°C, 960°C) on the magnetic and dielectric properties, as well as low-loss characteristics of $Mg_{0.7}Cd_{0.3}Fe_{1.92}Ga_{0.08}O4$ composites with 5 wt.% Bi_2O_3 addictive. The increase in permeability is due to an increase in magnetization (saturation magnetization Ms from 24.37 emu/g to 33.12 emu/g). The actual permeability (μ') of Mg-Cd-Ga ferrites was shown to increase in a monotonic manner from 15 to 19 H/m. Increasing temperature also fine-tuned the actual portion of permittivity (ε'). As a consequence, at 920°C, and are almost equal in frequency ranges extending from 1 MHz to, and these properties may be exploited for antenna miniaturisation and excellent radiation behaviour. Furthermore, the extremely low magnetic ($\tan\delta_\mu \sim 3*10^{-2}$) and dielectric ($\tan\delta_\varepsilon \sim 5*10^{-3}$) tangents allow for high radiation efficiency during operation. These findings showed that the suggested materials have a promising future as high-frequency antenna substrate.

The broad-band electromagnetic characteristics of composites of $(1-x)SrFe_{12}O_{19}$-xLi_2MoO_4, x = 0.4, 0.5, 0.6, and 0.7 with density up to 91%, were synthesized using a cold-sintering technique by Rajan et al. 2021 [52]. In the composites, X-ray diffraction (XRD) examination as shown in Fig 4 indicated the coexistence of $SrFe_{12}O_{19}$ (SFO) and Li_2MoO_4 (LMO) phases, with no other phases.

With an increase in the LMO volume fraction, the evolution of microstructure allowing increased densification was seen. The dielectric loss ($\tan\delta_\varepsilon$) dropped as the LMO volume percentage rose, but the real permittivity (ε') increased. Furthermore, all of the composites have a real permeability (μ') higher than unity, and the magnetic loss ($\tan\delta_\mu$) is on the scale of 10^{-2}. A ferrite resonator antenna (FRA) integrated with the SFOLMO composite was developed, modelled, and built to show the magneto dielectric composite's application potential in microwave antenna applications. The manufactured FRA resonating at 12.89 GHz has a very high return loss of -40 dB and a 510 MHz broad impedance bandwidth Fig. 5.

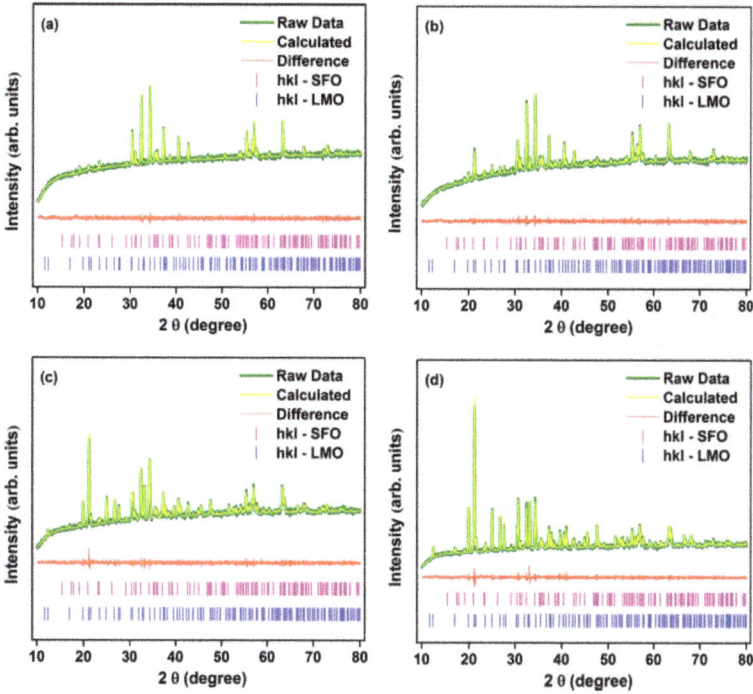

Figure 4. Refined X-ray diffraction (XRD) patterns of $(1-x)SrFe_{12}O_{19}-xLi_2MoO_4$ cold-sintered composites with (a) x = 0.4, (b) x = 0.5, (c) x = 0.6, and (d) x = 0.7. Reprinted (adapted) with permission from ref [52] Copyright {2021} American Chemical Society.

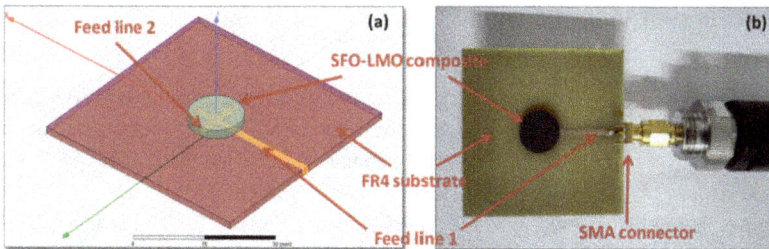

Figure 5. Geometry of the (a) simulated FRA and (b) fabricated FRA. Reprinted (adapted) with permission from ref [52] Copyright {2021} American Chemical Society.

The constructed ferrite resonator antenna's exceptional characteristics imply that it might be a good option for Ku-band (microwave part of the electromagnetic spectrum with frequencies ranging from 12 to 18 gigahertz (GHz) applications.

Rahman et al. used the sol-gel process to make $Co_xZn_{(0.90-x)}Al_{0.10}Fe_2O_4$ ferrites, which were used as flexible microwave substrates for microstrip patch antennas [53]. In addition, the suggested Co-Zn ferrites samples' dielectric and magnetic characteristics were tested for microwave application confirmation. The magnetic permeability ε_r and magnetic loss tangent (T_δ) values are also calculated to generate the specified scattering parameters S11 and S21 derived from rectangular substrate pellets, with μr and $(T_{\delta m})$ values ranging from 0.90 to 1.00 and 0.003 to 0.007, respectively. A modified diamond-shaped microstrip patch antenna is developed and produced on the suggested flexible substrates made from synthetic Co-Zn ferrites nanopowders mixed with PVA binder to demonstrate its use in the microwave regime's S-band.

In a research article Rajendran Lekshmi synthesized PMMA -NFO ($NiFe_2O_4$) nanoparticles composites with different volume fractions of NFO (0.05 V_f, 0.1 V_f, 0.15 V_f, 0.2 V_f) [54]. On the basis of result obtained, a remarkable miniaturization near about 95.46% was monitored for this Magneto Dielectric antenna as compared to normal dielectric substrates possess permittivity and permeability values equivalent to unity. Alteration in magnetocapacitance because of magnetic field is represented in Fig. 6(a). The fluctuation in magneto capacitance with increasing magnetic field is shown in this diagram. The magnetic and charge ordering in the magnetic phase of the composite will be reduced by a magnetic field, and the MC value will alter as a result [59]. As a result, at room temperature, an increase in the magnetic field will result in a higher negative magnetocapacitance value in the composite. However, no trace of an electromechanical resonance frequency can be found anywhere in the frequency range we looked at.

Because the MD characteristics show promise, they can be employed to benefit antenna shrinking in the future. Through a polymer hot pressing process, a functionally graded anisotropic substrate with attributes such as er and mr that change in the z plane (Fig. 6(b)).

A model assisted study on the fabrication of a microstrip patch antenna (MPA) is carried out to validate the applicability of the proposed laminar MD composites for size reduction in antennas. A comparison of the geometric parameters of the optimised designs of lossless dielectric substrate LD, ungraded magneto dielectric material UMD, and graded magneto dielectric material GMD substrate MPAs, as shown in Figures 7A and b, can be used to assess the efficiency of magnetodielectric loading in size reduction.

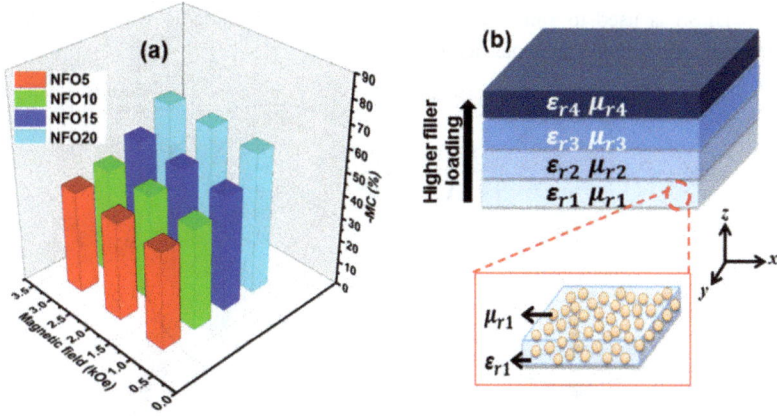

Figure 6. Variation in magneto capacitance with the application of different magnetic fields in the PMMA–NFO composites. (b) A schematic showing the developed functionally graded MD substrate and the enlarged single layer of the composite that contains magnetic filler in the PMMA matrix. Reprinted (adapted) from ref [54] Copyright {2022} Royal Society of Chemistry.

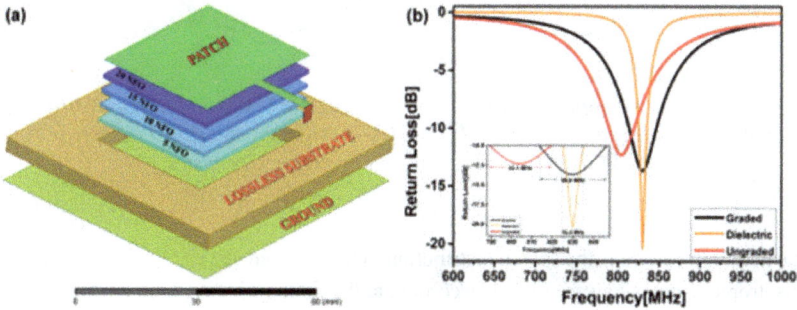

Figure 7. Schematic showing MPA employing the GMD substrate and the uniaxial variation in MD properties. (b) Return loss characteristics of MPAs with LD substrate, and GMD and UMD substrates. (b inset) Bandwidth comparison of MPAs with the LD substrate, and GMD and UMD substrate Reprinted (adapted) from ref [54] Copyright {2022} Royal Society of Chemistry.

ANSYS HFSS is used to simulate the spatial distribution of radio wave strength and antenna gain of modelled antennas, as shown in Fig. 8.

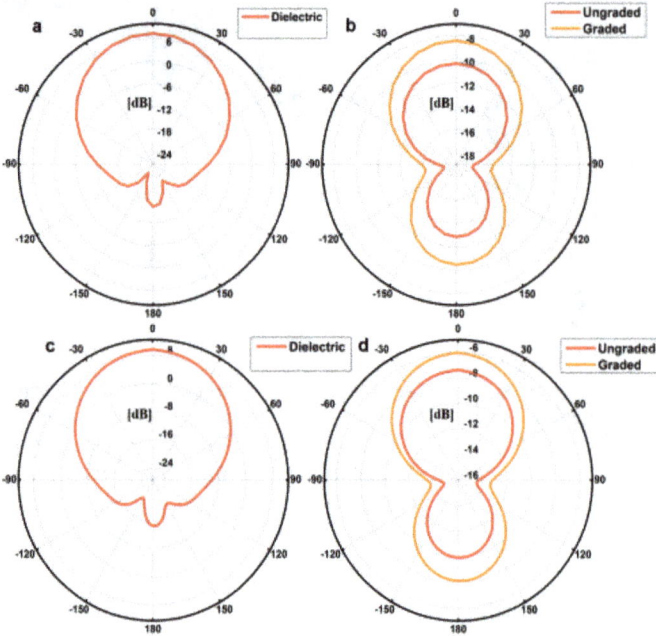

Figure 8. Radiation pattern of (a) LD substrate, and (b) GMD and UGD substrate; gain plots of (c) LD substrate, and (d) GMD and UMD substrate. Reprinted (adapted) from ref 54 Copyright {2022} Royal Society of Chemistry.

The radiation pattern is the spatial distribution of the radiated waves. A functionally graded anisotropic composite was created, according to the findings by stacking the various filler-loaded samples in a methodical manner, with laminates with a higher level of solid loading on the top layer the suggested strategy. Patch antennas with functionally gradient and axially anisotropic substrates are appropriate for large-scale antennas.

4.   Factors affecting the performance of antenna

4.1   Size

The HF antenna is quite large due to the enormous wavelength of the HF band. As a result, shrinking the HF antenna is a good idea. However, shrinking the antenna's size reduces its efficiency. Shih et al. proposed using a platform-mounted antenna to excite the platform's characteristic modes (CMs) in order to enhance overall efficiency and gain at lower frequencies [55]. The wavelength of the lower limit of the HF band is related to the platform. The effective size of the antenna is raised by stimulating a partial or entire size of the platform, which improves efficiency at lower frequencies. The platform's shape and size, however, have a significant impact on the design's performance. This technique is not applicable to all antennas and platforms. Microwave antennas are significantly smaller and need much less area than HF antennas. The antenna's size is no longer the most important issue. Other variables, such as dielectric material losses, become increasingly important.

4.2   Losses in dielectric material

Microstrip antennas are often made of low-cost, high-loss materials, resulting in low antenna efficiency. Chen et al. [56] proposed coupling arrangements to improve the efficiency of microstrip antennas. They created a non-resistive coupling structure antenna. The loss in the divider network is reduced without the resistors, and as a result, the efficiency is enhanced. The geometry of this type of antenna, on the other hand, is complicated and difficult to construct. Yang et al. proposed a planar-structured high-efficiency antenna [57]. The ground plane of the antenna is made of metasurface. Because the periodic structure has a low loss characteristic, the antenna's efficiency increases. Pan et al. [58] created an antenna with a metasurface as the patch. The frequency selective surface (FSS) was discovered to have filtering properties, and the metasurface inherited this characteristic. As a result of the metasurface patch's reduced insertion loss, the antenna's efficiency can be improved. The antenna's periodic structure can improve its efficiency [59].

4.3   The loss in propagation

The elements that influence the performance of millimetre wave antennas differ from those that influence the performance of HF and microwave antennas. The propagation loss in millimetre wave communication systems is substantial, for example, 15-30 dB/km at 60 GHz [60]. As a result, antenna arrays with excellent directivity and efficiency are required. The loss in the antenna array's feed network becomes the most important element impacting the array's overall efficiency, thus it's critical to maintain the loss in the feed network as

An Introduction to Hard Ferrites: From Fundamentals to Practical Applications   Materials Research Forum LLC
Materials Research Foundations **142** (2023) 152-184   https://doi.org/10.21741/9781644902318-6

low as possible. Zarifi et al. proposed a slot antenna array with excellent efficiency [61]. The antenna is built using the waveguide gap method, which eliminates the requirement for electrical contact between the layers. This antenna, on the other hand, is complicated, with three layers. It is difficult to create. Wu et al. [62] developed a high-efficiency millimetre wave antenna that can be manufactured on a soft substrate. An array of patch antennas is proposed as the antenna. Slots feed the patches, avoiding the necessity of high-loss millimetre wave connections. Microstrip lines are utilised in the feed network. Because the width of the microstrip line and the thickness of the substrate are both quite big, there is a significant amount of loss in the feed network. The feed network was designed by Park et al using a substrate integrated waveguide (SIW) [63]. The design of SIW and gap coupling techniques are required approaches in order to build a high-efficiency and easy-to-fabricate millimetre wave antenna.

4.4   Return loss

A horizontally polarised dipole antenna must be positioned at a height of at least $\lambda/2$ to assure the antenna's impedance and radiation performance [64]. The influence of the image current cannot be ignored when the current flowing through the antenna is situated near the ground plane [65]. As illustrated in Fig. 9, the image current on the ground plane is in the opposite direction of the current in the antenna. The ground plane reflects the electromagnetic wave emitted by the antenna. The reflected wave shifts 180 degrees in phase. As a result, the wave from the image current cancels the reflected wave, resulting in the cancellation of radiated power by the image current.

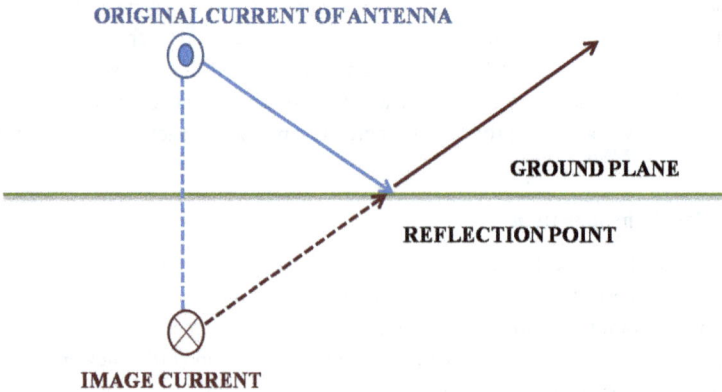

Figure 9. Image current direction.

The authors used ferrite materials to construct a broadband (3 MHz - 6 MHz) bow-tie shaped HF antenna that is near the ground plane with a very low height and eliminates the destructive influence of the image current [66]. The influence of image current cannot be ignored when the antenna is close to a conductive ground plane. The authors utilised commercial ferrite tiles to cover the whole conductive ground to decrease the influence of image current and mutual resistance. It can be observed that a significant quantity of commercial ferrites may be utilised to enhance the voltage standing wave ratio of the near ground antenna. The overall efficiency, however, is not displayed. Even if the return loss is decreased, the antenna's efficiency is poor due to the lossy ferrite location.

4.5   Radiation efficiency

Zhao et al. suggested broadband monopoles for VHF-UHF band applications [67, 68]. However, the antenna's overall efficiency was still too low. At higher frequencies, the ferrite loading has no effect on the antenna's realised gain. At higher frequencies, the realised gain is even reduced significantly. The use of ferrite loading does not increase the total volume of the antenna, but it does substantially increase the weight of the antenna due to the high weight of the ferrite utilised. The arrangement of the commercial ferrite bars needs to be improved in order to reduce the antenna's weight. Moon et al. employed four distinct loading ferrite bars with varying heights and widths to reduce the antenna's weight while retaining its performance [69]. The authors recommended placing the ferrite in locations with large magnetic fields to reduce the required ferrite even further. The loading ferrite bars were kept to a bare minimum. The voltage standing wave ratio is considerably enhanced as compared to the antenna without ferrite loading. The loop antenna's major source of energy is magnetic.

The loading ferrite has a high permeability characteristic. As a result, the antenna's energy storage decreases, and the Q factor decreases [70]. The realised gain is improved at lower frequencies but lowered at higher frequencies when compared to the antenna without ferrite loading. The gain is nearly constant at lower frequencies and somewhat enhanced at higher frequencies as compared to the antenna with one full piece of ferrite. Because ferrite is used less at higher frequencies, the realised gain improves. The return loss is considerably decreased at lower frequencies. There will be less energy reflected. As a result, overall efficiency and realised gain both improve.

5.   Artificial materials to improve efficiency

The ferrite is useful in improving the antenna's performance in the HF band. Due to its unique EM characteristics, the periodic structure is useful for improving antenna efficiency when the frequency rises to higher bands, such as the microwave band. Two of the most essential characteristics are permittivity and permeability. The permittivity (ε) and permeability (μ) can be used to characterise the EM characteristics of the substrate while designing a microstrip antenna. The most common substance in nature is a double positive medium, which has permittivity and permeability both greater than zero ($\varepsilon > 0$, and $\mu > 0$). Artificial materials arise in various applications because they require materials with non-naturally occurring or specialised characteristics [71]. By embedding different elements with unique shapes in some host medium, an artificial material based on periodic structure may be created. A vast number of independent parameters are provided by the periodic structure. Designers can obtain materials with specific permittivity and permeability because to the great degree of flexibility.

5.1   Use of substrate integrated waveguide (SIW) to reduce loss

Millimeter wave antennas are quickly evolving in tandem with the development of 5G communication methods. Because of the significant transmission loss in the air, a high realised gain is one of the most essential criteria. The antenna array is suggested in the millimetre wave communication system to obtain high gain. The feed network is widely recognised to be the primary source of loss in a big antenna array. As a result, in order to reduce losses in the antenna array's feed network, it is important to examine the performance of commonly used transmission techniques. The most common EM transmission techniques are rectangular waveguide, coaxial line, and microstrip line. Fig. 10 illustrates their geometries.

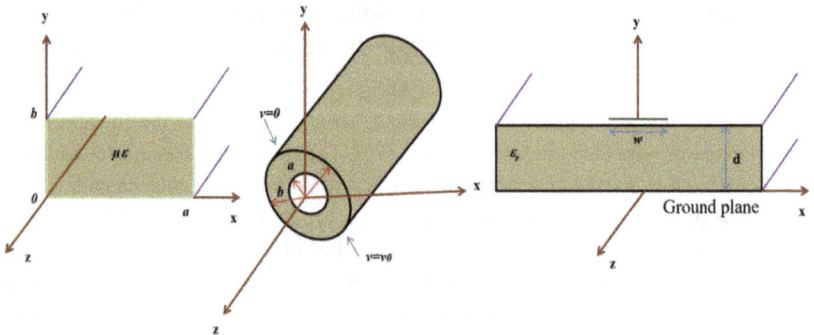

Figure 10. Structure of rectangular waveguide, coaxial line, and microstrip line for transmission

## 6.	Future prospects of antenna

Self-adaptation and self-improvement are the future prospects of reconfigurable antennas, with the goal of achieving maximum energy efficiency with minimal losses and low power requirements, in order to maintain the communication link with the highest reliability in a dynamic and unpredictable environment. Future reconfigurable antennas should be capable of performing many roles at the same time with no noticeable hiccups. In order to perceive and adapt to dynamic RF changes, reconfiguration approaches should be increasingly oriented towards developing fields like software driven IoT gateways or machine learning models linked to ANN technology. New wireless technology developments, such as cognitive radio systems, developing cellular systems, and so on, should be used as a foundation for the development of reconfigurable antennas.

Conclusion

To begin with, the most important affecting elements were observed in various frequency bands. In the microwave band, the dielectric loss of the substrate is the most important element affecting the microstrip antenna's efficiency. The loss in the feed network is the major loss component that lowers the efficiency of a millimetre wave antenna array. Ferrites have been proposed in the HF band to decrease the impact of near-ground antenna image current. The HF antenna's mismatch loss and efficiency can be enhanced. Because the amount and strength of the image current is determined by the current density on the ground, the use of ferrites is more successful when the ferrites are positioned in a region with a high current density. The ferrites can also be utilised to enhance the bandwidth of the near ground HF antenna while reducing its size. The bottom bound of the bandwidth can be increased by adding ferrites. When the periodic structures are placed in the material, the characteristics of the original material will alter. The permittivity is lowered and the permeability is raised in a specific frequency range by etching periodic patterns in the ground plane. The radiator can also be made out of the periodic structure. With the periodic structures in the antenna patch, more energy can be radiated out, and the antenna's bandwidth may be substantially expanded, according to the studies of the transmission line model and the dispersion relation.

This book chapter concluded method used to reconfigure antenna with hard ferrites. Effect of increasing the relative permeability of the substrate material on the physical size and efficiency of a microstrip antenna. Although using a high-permittivity substrate helps minimise device size, it also reduces gain, radiation pattern, and bandwidth. Using magneto-dielectric materials, it appears that the size of an antenna may be reduced without compromising bandwidth. This happens due to the field's confinement in a high-permittivity region and an impedance mismatch with the surrounding medium. It was also concluded that ferrite materials have a bright future as high-frequency antenna substrates.

References

[1] A.A.Mills, The lodestone: History, physics, and formation, Ann. Sci. 61 (2004), 273-319. https://doi.org/10.1080/00033790310001642812

[2] W. Gilbert, De magnete, Courier Corporation,1958, pp. 1-368

[3] C.B.Carter, M.G.Norton, Ceramic materials: science and engineering, Springer Science & Business Media, 2007, pp. 712

[4] A. Houbi, Z.A. Aldashevich, Y. Atassi, Z.B. Telmanovna, M. Saule, K. Kubanych, Microwave absorbing properties of ferrites and their composites: a review, J. Magn. Magn. Mater. 529 (2021) 167839. https://doi.org/10.1016/j.jmmm.2021.167839

[5] S. Zhang, W.E. Lee, Spinel-containing refractories, Mechanical engineering-New York and basel-marcel dekker then crc press/taylor and francis, 2004 pp. 215. https://doi.org/10.1201/9780203026328.ch9

[6] W. Bragg, X-rays and Crystal Structure, The Sci. Mon. 20 (1925) 115-121. https://doi.org/10.1038/115266a0

[7] S. Nishikawa, Structure of some crystals of spinel group, Proceedings of the Tokyo Mathematico-Physical Society. 2nd Series, 8 (1915) 199-209.

[8] M.K. Warshi, V. Mishra, A. Sagdeo, V. Mishra, R. Kumar, P.R. Sagdeo, Structural, optical and electronic properties of RFeO3, Ceram. Int. 44 (2018) 8344-8349. https://doi.org/10.1016/j.ceramint.2018.02.023

[9] B. Kim, H. Rhyu, I.Y. Lee, J. Byun, B. Lee, Compact internal antenna using a ferrite material for DVB-H reception in mobile phones, IEEE Antennas Propag. Soc. (2008) 1-4.

[10] R.C. Hansen, M. Burke, Antennas with magneto dielectrics, Microw. Opt. Technol. Lett. 26 (2000) 75-78. https://doi.org/10.1002/1098-2760(20000720)26:2<75::AID-MOP3>3.0.CO;2-W

[11] H. Mosallaei, K. Sarabandi, Magneto-dielectrics in electromagnetics: Concept and applications, IEEE Trans. Antennas Propag. 52 (2004) 1558-1567. https://doi.org/10.1109/TAP.2004.829413

[12] J. Costantine, Y. Tawk, S.E. Barbin, C.G. Christodoulou, Reconfigurable antennas: Design and applications, Proceedings of the IEEE, 103 (2015) 424-437. https://doi.org/10.1109/JPROC.2015.2396000

[13] D.H. Schaubert, F.G. Farrar, S.T. Hayes, A.R. Sindoris, US Department of Army, Frequency-agile, polarization diverse microstrip antennas and frequency scanned arrays. U.S. Patent 4, 367, 474. (1983).

[14] C.G.Christodoulou, Y. Tawk, S.A. Lane, S.R. Erwin, Reconfigurable antennas for wireless and space applications, Proceedings of the IEEE, 100 (2012), 2250-2261. https://doi.org/10.1109/JPROC.2012.2188249

[15] K. Entesari, A.P. Saghati, Fluidics in microwave components, IEEE Microw. Mag. 17 (2016) 50-75. https://doi.org/10.1109/MMM.2016.2538513

[16] Kojima, H., Handbook of Ferromagnetic Materials, Fundamental properties of hexagonal ferrites with magnetoplumbite structure, 1982, pp.305-391. https://doi.org/10.1016/S1574-9304(05)80091-4

[17] J. Jeong, K.W. Cho, D.W. Hahn, B.C. Moon, Y.H. Han, Synthesis of Co2Z Ba-ferrites, Mater. Lett. 59 (2005) 3959-3962. https://doi.org/10.1016/j.matlet.2005.07.044

[18] J.S. Kim, Y.H. Lee, B. Lee, J.C. Lee, J.J. Choi, J.Y. Kim, Effects of magneto-dielectric ceramics for small antenna application, J. Electr. Eng. Technol. 9 (2014) 273-279. https://doi.org/10.5370/JEET.2014.9.1.273

[19] S.M. Wiederhorn, Fracture surface energy of glass, J. Am. Ceram. Soc. 52 (1969) 99-105. https://doi.org/10.1111/j.1151-2916.1969.tb13350.x

[20] E. Neckenburger, H. Severin, J.K. Vogel, G. Winkler, Ferrite hexagonaler Kristallstrustur mit hoher Grenzfrequenz, Z Angew. Phys. 18 (1964) 65.

[21] M.A.Vinnik, Phase relationships in the bao-coo-fe_2o_3 system, Russ. J. Inorg. Chem. 10 (1965) 1164-1167.

[22] S.I. Kuznetsova, E.P. Naiden, T.N. Stepanova, Topotactic reaction kinetics in the formation of the hexagonal ferrite Ba3Co2Fe24O41, Inorg. Mater. 24 (1988) 856-859.

[23] J. Drobek, W.C. Bigelow, R.G. Wells, Electron microscopic studies of growth structures in hexagonal ferrites, J. Am. Ceram. Soc. 44 (1961) 262-264. https://doi.org/10.1111/j.1151-2916.1961.tb15375.x

[24] M.A. Ahmed, N. Okasha, S.I. El-Dek, Preparation and characterization of nanometricMn ferrite via different methods, Nanotechnology, 19 (2008) 065603. https://doi.org/10.1088/0957-4484/19/6/065603

[25] J. Dufour, E. López-Vidriero, C. Negro, R. Latorre, E.M. Alcala, F. López-Mateos, A. Formoso, Improvement of ceramic method for synthesizing M-type hexaferrites, Chem. Eng. Commun. 167 (1998) 227-244 https://doi.org/10.1080/00986449808912702

[26] R.K. Tenzer, Influence of particle size on the coercive force of barium ferrite powders, J. Appl. Phys. 34 (1963) 1267-1268. https://doi.org/10.1063/1.1729465

[27] H.F. Yu, K.C. Huang, Preparation and characterization of ester-derived BaFe12O19 powder. J. Mater. Res. 17 (2002) 199-203. https://doi.org/10.1557/JMR.2002.0029

[28] C.D. Mee, J.C. Jeschke, Single-domain properties in hexagonal ferrites, J. Appl. Phys. 34 (1963) 1271-1272. https://doi.org/10.1063/1.1729467

[29] W. Roos, Formation of chemically coprecipitated barium ferrite, J. Am. Ceram. Soc. 63 (1980) 601-603. https://doi.org/10.1111/j.1151-2916.1980.tb09843.x

[30] G. Xiong, M. Xu, Z. Mai, Magnetic properties of Ba4Co2Fe36O60 nanocrystals prepared through a sol-gel method, Solid State Commun. 118 (2001) 53-58. https://doi.org/10.1016/S0038-1098(01)00031-X

[31] S. Kour, R.K. Sharma, R. Jasrotia, V.P.Singh, A brief review on the synthesis of maghemite (γ-Fe2O3) for medical diagnostic and solar energy applications, AIP Conference Proceedings (2019), 090007. https://doi.org/10.1063/1.5122451

[32] S.K. Mishra, L.C. Pathak, V. Rao, Synthesis of submicron Ba-hexaferrite powder by a self-propagating chemical decomposition process, Mater. Lett. 32 (1997) 137-141. https://doi.org/10.1016/S0167-577X(97)00027-X

[33] Y.S. Hong, C.M. Ho, H.Y. Hsu, C.T. Liu, Synthesis of nanocrystalline Ba (MnTi) xFe12− 2xO19 powders by the sol-gel combustion method in citrate acid-metal nitrates system (x= 0, 0.5, 1.0, 1.5, 2.0), J. Magn. Magn. Mater. 279 (2004) 401-410. https://doi.org/10.1016/j.jmmm.2004.02.008

[34] L. Junliang, Z. Yanwei, G. Cuijing, Z. Wei, Y. Xiaowei, One-step synthesis of barium hexaferritenano-powders via microwave-assisted sol-gel auto-combustion, J. Eur. Ceram. Soc. 30 (2010) 993-997. https://doi.org/10.1016/j.jeurceramsoc.2009.10.019

[35] Z. Lalegani, A. Nemati, Influence of synthesis variables on the properties of barium hexaferrite nanoparticles, J. Mater. Sci.: Mater. Electron. 28 (2017) 4606-4612. https://doi.org/10.1007/s10854-016-6098-5

[36] V. Pillai, P. Kumar, M.J. Hou, P. Ayyub, D.O. Shah, Preparation of nanoparticles of silver halides, superconductors and magnetic materials using water-in-oil microemulsions as nano-reactors, Adv. Colloid Interface Sci. 55 (1995) 241-269. https://doi.org/10.1016/0001-8686(94)00227-4

[37] R.R. Schaller, Moore's law: past, present and future, IEEE spectrum, 34 (1997) 52-59. https://doi.org/10.1109/6.591665

[38] R.C. Hansen, Fundamental limitations in antennas, Proceedings of the IEEE, 69 (1981) 170-182. https://doi.org/10.1109/PROC.1981.11950

[39] K. Carver, J. Mink, Microstrip antenna technology, IEEE Trans. Antennas Propag. 29 (1981) 2-24. https://doi.org/10.1109/TAP.1981.1142523

[40] F. Kuroki, Y.S. Takigawa, S. Kashihara, Radiation Characteristics of Integrated Traveling-Wave Antenna Etched on Heavily-High Permittivity Substrate for Size Reduction, IEEE Radio and Wireless Symposium (2007) 169-172. https://doi.org/10.1109/RWS.2007.351794

[41] A. Saini, A. Thakur, P. Thakur, Matching permeability and permittivity of $Ni0.5Zn0.3Co0.2In0.1Fe1.9O4$ ferrite for substrate of large bandwidth miniaturized antenna, J. Mater. Sci.: Mater. Electron. 27 (2016) 2816-2823. https://doi.org/10.1007/s10854-015-4095-8

[42] A. Saini, A. Thakur, P. Thakur, Effective permeability and miniaturization estimation of ferrite-loaded microstrip patch antenna, J. Electron. Mater. 45 (2016) 4162-4170. https://doi.org/10.1007/s11664-016-4634-y

[43] P. Ikonen, S. Tretyakov, On the advantages of magnetic materials in microstrip antenna miniaturization, Microw. Opt. Technol. Lett. 50 (2008) 3131-3134. https://doi.org/10.1002/mop.23931

[44] K. Borah, N.S. Bhattacharyya, Magnetodielectric composite with ferrite inclusions as substrates for microstrip patch antennas at microwave frequencies, Compos. B. Eng. 43 (2012) 1309-1314. https://doi.org/10.1016/j.compositesb.2011.11.067

[45] Z. Zheng, H. Zhang, J.Q. Xiao, Q. Yang, L. Jia, Low loss NiZn spinel ferrite-W-type hexaferrite composites from BaM addition for antenna applications, J. Phys. D: Appl. Phys. 47 (2014) 115001. https://doi.org/10.1088/0022-3727/47/11/115001

[46] J.L. Mattei, E. Le Guen, A. Chevalier, A.C. Tarot, Experimental determination of magnetocrystalline anisotropy constants and saturation magnetostriction constants of NiZn and NiZnCo ferrites intended to be used for antennas miniaturization, Journal of Magnetism and Magnetic Materials, 374 (2015) 762-768. https://doi.org/10.1016/j.jmmm.2014.09.026

[47] A. Saini, A. Thakur, P. Thakur, Effective permeability and miniaturization estimation of ferrite-loaded microstrip patch antenna, J. Electron. Mater. 45 (2016) 4162-4170 https://doi.org/10.1007/s11664-016-4634-y

[48] A. Saini, A. Thakur, P. Thakur, Miniaturization and bandwidth enhancement of a microstrip patch antenna using magneto-dielectric materials for proximity fuze application, Journal of Electronic Materials, 46 (2017) 1902-1907. https://doi.org/10.1007/s11664-016-5256-0

[49] K. Polley, T. Alam, J. Bera, Synthesis and characterization of BaFe12O19-CoFe2O4 ferrite composite for high-frequency antenna application, J. Aust. Ceram. Soc. 56 (2020) 1179-1186. https://doi.org/10.1007/s41779-020-00477-x

[50] O. Alcalá, S. Briceño, W. Brämer-Escamilla, P. Silva, Toroidal cores of MnxCo1−xFe2O4/PAA nanocomposites with potential applications in antennas, Materials Chemistry and Physics, 192 (2017) 17-21. https://doi.org/10.1016/j.matchemphys.2017.01.035

[51] G. Gan, D. Zhang, J. Li, G. Wang, Y. Yang, X. Wang, H. Zhang, Effect of temperature on magnetic and dielectric properties of Mg-Cd-Ga ferrites for high-frequency-range antennas. In Journal of Physics: Conference Series 1802 (2021) 022078. https://doi.org/10.1088/1742-6596/1802/2/022078

[52] A. Rajan, S.K. Solaman, S. Ganesanpotti, Cold Sintering: An Energy-Efficient Process for the Development of SrFe12O19-Li2MoO4 Composite-Based Wide-Bandwidth Ferrite Resonator Antenna for Ku-Band Applications, ACS Appl. Electron. Mater. 3.5 (2021) 2297-2308. https://doi.org/10.1021/acsaelm.1c00196

[53] M.A. Rahman, M.T. Islam, M.J. Singh, I. Hossain, H. Rmili, M. Samsuzzaman, Magnetic, dielectric and structural properties of CoxZn (0.90-x) Al0. 10Fe2O4 synthesized by sol-gel method with application as flexible microwave substrates for microstrip patch antenna, J. Mater. Res. Technol. 16 (2022) 934-943. https://doi.org/10.1016/j.jmrt.2021.12.058

[54] D.R. Lekshmi, S.P. Adarsh, M. Bayal, S.S. Nair, K.P. Surendran, Functionally graded magnetodielectric composite substrates for massive miniaturization of

microstrip antennas, Mater. Adv. 3 (2022) 2380-2392.
https://doi.org/10.1039/D1MA00844G

[55] T.Y. Shih, N. Behdad, Bandwidth enhancement of platform-mounted HF antennas using the characteristic mode theory, IEEE Trans. Antennas Propag. 64 (2016) 2648-2659. https://doi.org/10.1109/TAP.2016.2543778

[56] X. Chen, L. Yang, J.Y. Zhao, G. Fu, High-efficiency compact circularly polarized microstrip antenna with wide beamwidth for airborne communication, IEEE Antennas Wirel. Propag. Lett. 15 (2016) 1518-1521.
https://doi.org/10.1109/LAWP.2016.2517068

[57] W. Yang, D. Chen, W. Che, High-efficiency high-isolation dual-orthogonally polarized patch antennas using nonperiodic RAMC structure. IEEE Trans. Antennas Propag. 65 (2016) 887-892. https://doi.org/10.1109/TAP.2016.2632700

[58] Y. M. Pan, P.F. Hu, X.Y. Zhang, S.Y Zheng, A low-profile high-gain and wideband filtering antenna with metasurface, IEEE Trans. Antennas Propag. 64 (2016) 2010-2016. https://doi.org/10.1109/TAP.2016.2535498

[59] G. Yang, J. Li, R. Xu, Y. Ma, Y. Qi, Improving the performance of wide-angle scanning array antenna with a high-impedance periodic structure, IEEE Antennas Wirel. Propag. Lett. 15 (2016)1819-1822.
https://doi.org/10.1109/LAWP.2016.2537850

[60] R.C. Daniels, R.W. Heath, 60 GHz wireless communications: Emerging requirements and design recommendations, IEEE Veh. Technol. Mag., 2 (2007) 41-50.
https://doi.org/10.1109/MVT.2008.915320

[61] D. Zarifi, A. Farahbakhsh, A.U. Zaman, P.S. Kildal, Design and fabrication of a high-gain 60-GHz corrugated slot antenna array with ridge gap waveguide distribution layer, IEEE Trans. Antennas Propag. 64 (2016) 2905-2913.
https://doi.org/10.1109/TAP.2016.2565682

[62] J. Wu, Y.J. Cheng, Y. Fan, Millimeter-wave wideband high-efficiency circularly polarized planar array antenna, IEEE Trans. Antennas Propag. 64 (2015) 535-542.
https://doi.org/10.1109/TAP.2015.2506726

[63] S.J. Park, S.O. Park, LHCP and RHCP substrate integrated waveguide antenna arrays for millimeter-wave applications, IEEE Antennas Wirel. Propag. Lett. 16 (2016) 601-604. https://doi.org/10.1109/LAWP.2016.2594081

[64] A.A.Lestari, A.G. Yarovoy, L.P.Ligthart, Ground influence on the input impedance of transient dipole and bow-tie antennas, IEEE Trans. Antennas Propag. 52 (2004) 1970-1975. https://doi.org/10.1109/TAP.2004.832371

[65] J.D. Kraus, R.J. Marhefka, A.S. Khan, Antennas and wave propagation, Tata McGraw-Hill Education, 2006.

[66] W. Lin, Z. Shen, Broadband horizontally polarized HF antenna with extremely low profile above conducting ground. In 2013 IEEE Antennas and Propagation Society International Symposium (APSURSI) 2013 (pp. 688-689). IEEE. https://doi.org/10.1109/APS.2013.6711004

[67] J. Zhao, C.C. Chen, J.L.Volakis, Frequency-scaled UWB inverted-hat antenna, IEEE Trans. Antennas Propag, 58 (2010) 2447-2451. https://doi.org/10.1109/TAP.2010.2048866

[68] J. Zhao, T. Peng, C.C. Chen, J.L Volakis, Low-profile ultra-wideband inverted-hat monopole antenna for 50 MHz-2 GHz operation, Electron. Lett. 45 (2009) 142-144. https://doi.org/10.1049/el:20092571

[69] H. Moon, G.Y. Lee, C.C. Chen, J.L.Volakis, An extremely low-profile ferrite-loaded wideband VHF antenna design, IEEE Antennas Wirel. Propag. Lett. 11(2012) 322-325. https://doi.org/10.1109/LAWP.2012.2191131

[70] L.J. Chu, Physical limitations of omni-directional antennas, J. Appl. Phys., 19(1948) 1163-1175. https://doi.org/10.1063/1.1715038

[71] N. Engheta, R.W. Ziolkowski, Metamaterials: Physics and Engineering Explorations, Hoboken/Piscataway, 2006. https://doi.org/10.1002/0471784192

An Introduction to Hard Ferrites: From Fundamentals to Practical Applications   Materials Research Forum LLC
Materials Research Foundations **142** (2023) 185-206   https://doi.org/10.21741/9781644902318-7

Chapter 7

Applications of Hard Ferrites in Memory Devices

Ritesh Verma[1,2], Ankush Chauhan[3], Rajesh Kumar[4*]

[1]Department of Physics, Amity University, Haryana, Gurugram-122413, India

[2]Himalayan Centre of Excellence in Nanotechnology, Shoolini University of Biotechnology and Management Sciences, Bajhol, Solan (H.P.) India-173229

[3]Chettinad Acadamy of Research and Education, Kelambakkam, Kanchipuram, Tamil Nadu, India-603103

[4]Department of Physics, Sardar Patel University, Mandi (H.P.) India-175001

*rajesh.shoolini@gmail.com

Abstract

Ferrites are the important material for memory devices. In this we discuss the ferrites in detail with their classification, preparation method to their applications. We also present a brief introduction about hard ferrites and their application in memory devices. Ferrites are the materials that offer distinct electrical and magnetic features that are helpful for various applications. It is noted that spin transmission torques may change magnetization through the current travelling through a magnetic tunnel interface, an effect followed by the spin transfer torque magnet random access memory as the switching mechanism. Also, it is observed that the transistor-type memory devices that employ nanostructured materials as loading sites for the trap are nano-floatting Gate (NFGM). Thus, this chapter presents a way forward for the memory devices.

Keywords

Ferrites, EMI Shielding, Spin Transmission, Magnetization, Nano-Floating Gate

Contents

1.   Introduction

For many hundred years, the history and use of ferrites (magnetic oxides) has been documented. A natural non-metallic solid, loadstone (magnetite, Fe_3O_4) may attract iron, which was recorded in ancient Greek documents about 800 B.C first. Much later magnetite was utilized as 'Lodestones,' which was used to find magnet North by early navigators. After the first technical magnetic substance, this is the first scientific importance, since it created the first compass. In 1600 William Gilbert published the first scientific studies on magnetism called De Magnete. Later Hans Christian Oersted noticed in 1819 that a magnetic compass needle was influenced by an electric current in a wire. Magnetite is of course a weak 'hard' ferrite. The 'hard' ferrites are a mostly permanent magnetic. Originally produced for inductors and antennas, "soft" ferrite expanded into innumerable sizes and forms for a wide range of purposes in a couple of selected formats and sizes. Ferries are also used mostly in three areas: low-level applications, electrical applications and the reduction of electro-magnetic interference (EMI). Ferries are still being applied more widely in electrical circuits. Ferrite components are the option for both traditional and creative uses because of a large variety of potential geometries, the continuous development in properties of materials and their relative economic productivity. Ferrites are fundamentally porcelain, dark grey or black, and extremely tough and brittle. Ferrites may be considered magnetic materials comprised of ferric ion containers (the name ferrite originates from the Latin "iron ferrum"), and classed as ferrimagnetic materials. Ferrite may be produced by high-temperature reaction method, sol–gel process, co-precipitation, pulsed laser deposition, powerful ball milling, and hydraulic technology in powder or thin film formats. To acquire the appropriate form and subsequently to convert it to a ceramic component by sintering, a ferrite core is created by pressing a blend of powder comprising the constituent raw components. The magnetic features result from interactions between metal ions that occupy certain locations in relation to the oxygen ions in the oxide's crystal structure.

2.   Classification of ferrites

Depending upon the crystal structure, ferrites are of following types.

- Spinel ferrites
- Garnet ferrites
- Ortho ferrites
- Hexagonal ferrites

2.1   Spinel ferrites

The chemical formula MFe_2O_4, where M stands for divalent metal ions, describes spinel ferrites, as shown in Fig. 1. Spinel ferrite's crystal structure has two interstitial sites, Tetrahedral (A) and Octahedral (B) [1]. Tetrahedral A site and octahedral B site accommodate a number of cations, making it possible for ferrites to vary widely. M may be substituted by different divalent metal ions and a number of spinel ferrites are possible. Other trivalent ions such as Al^{3+}, Cr^{3+}, Ga^{3+} and more can be substituted by Fe^{3+}. A mixture of divalent and tetravalent ions can also be substituted to Fe^{3+} ions.

2.2   Garnet ferrites

The ferrimagnetic garnet chemical formula is $Me_3Fe_5O_{12}$, where Me is a trivalent ion, as a rare earth or yttrium [3, 4]. Figure 2 presents the crystal structure of garnet ferrite. The cell of the unit consists of eight $Me_3Fe_5O_{12}$ molecules, i.e. (160 atoms). The metal ions are spread over three locations. The Me ions are located in the dodecahedral sites of eight oxygen ions, which are Fe^{3+} ions spread in ratio 3:2 over tetrahedral and octahedral sites. They are termed c-sites. The magnetic alignment is a result of super exchange through the intermediary ions for oxygen, as in Spinel ferrites, and with a shorter Me-O distance and a closer Me-O-Me angle to 180°, the interaction is predicted to be stronger. On this premise, the interaction from d to cations (both negative) is concluded reasonably strong. The magnetic alignments in the solid govern these interacts.

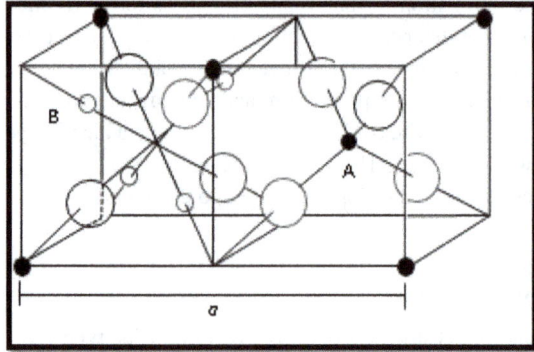

Figure 1. Diagram presents the unit cell for spinel ferrites [2].

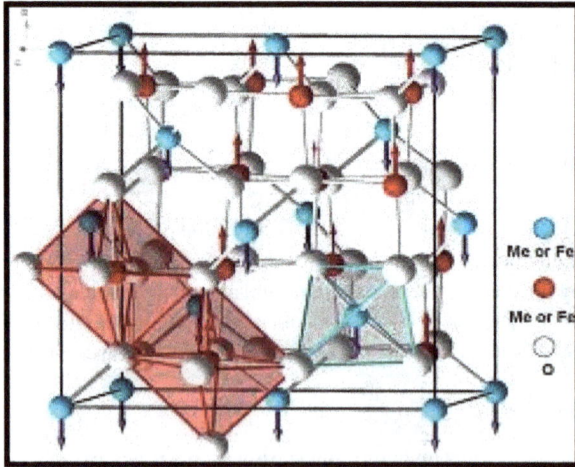

Figure 2. Image of garnet ferrite [5].

2.3   Ortho ferrites

The general formula $MeFeO_3$ is for the ortho-ferrite, whereby Me is a trivalent metal ion with large radius, such as rare-earth ion or Y. They crystallize using the orthorhombic unit cell in a deformed perovskite structure. The structure of orthoferrites is same as that of perovskite material, as shown in Fig.3.

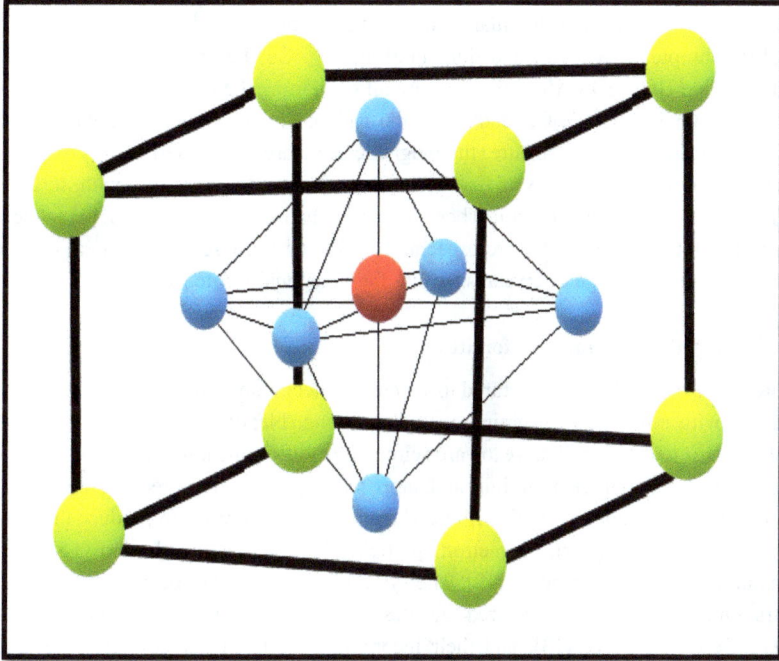

Figure 3. Presents the structure of orthoferrites unit cell [6]. "Reprinted with Permission from IOP, ECS Solid State Science and technology, 10 (2021) 073004."

These ortho-ferrites display mild ferromagnetism, due to the slight canting in aligning two lattices with anti-ferromagnetism. The angle is in order of ten to two radius but is enough to enter the antiferromagnetic axis in a tiny net ferromagnetic moment. At ambient temperature and parallel to the (100) axis, the rotary direction for Fe ion in $HOFeO_3$ was experimentally measured in both $HOFeO_3$ and $ErFeO_3$ and at 1.25K the direction for $HOFeO_3$ and $ErFeO_4$ was determined in $HOFeO_3$ (110). At a considerably lower Neel template, [6.5 K for $HOFeO_3$ and 4.3 K for $ErFeO_3$], the turning moment on the rare earth ion is ordered.

2.4 Hexagonal ferrites

Hexaferrites are a class of chemicals possessing hexagonal and rhombohedral symmetry, rather than a conventional crystal construction. In the upper portion of the $MeO-Fe_2O_3$-BaO ternary phase diagram the major compositions can be depicted, as shown in Fig.4. Me, for example, is a divalent cation, like Ni, Mg, Co, Fe, Zn, Cu. The $BaFe_{12}O_{19}$-$Me_2Fe_4O_8$ and $BaFe_{12}O_{19}$-$Me_2BaFe_{12}O_{22}$ are all containing the magnetic hexaferrites [7]. If $M(BaFe_{12}O_{19})$, $S(Me_2Fe_4O_8)$ and Y are the ending members ($Me_2BaFe_{12}O_{22}$). Due to their complicated crystal structure, the magnetic structure of hexaferrites is complex. In $BaFe_{12}O_{19}$, iron ions occupy five sub-lattices at low temperature, with a 20 µB magnetization per formulation unit. There is a comprehensive examination of magnetic and crystal structures. Hexaferrites are especially intriguing because of their great coercion. Figure 4 shows the image of crystal structure for hexagonal ferrite [8].

3.   Preparation methods for ferrites

Ferrites were originally manufactured using ceramic techniques for the production of bulk materials with micrometrical grains, including frying, blending, pressing, sintering and finishing. Due to the overall drive towards circuit integration and miniaturization, however, ferrite is produced in thick, thin films and, more recently, nanostructured. Ferrite thin films may be epitaxial or polycrystalline films. The most important ways to produce thin ferrite films are electroplating [9], magnetron sputtering [10], the laser pulsation [11] and the epitaxial of the molecular beam [12]. The synthesis, in particular, of heterostructures, of isostructural and quasi structured ferrites such as Fe_3O_4/NiO, Fe_3O_4/CoO, and $(Mn,Zn)Fe_2O_4/Co\ Fe_2O_4$ [13], and their incorporation into planar devices, is an added avenue to adjusting ferrite characteristics. As stated below, the combining of ferrite layers with piezoelectric layers leads to new and intriguing applications. New and technical advantages have been achieved by decreasing the scale to nanometric dimensions. New and technologically intriguing characteristics were achieved by decreasing the scale to the nanometric level.

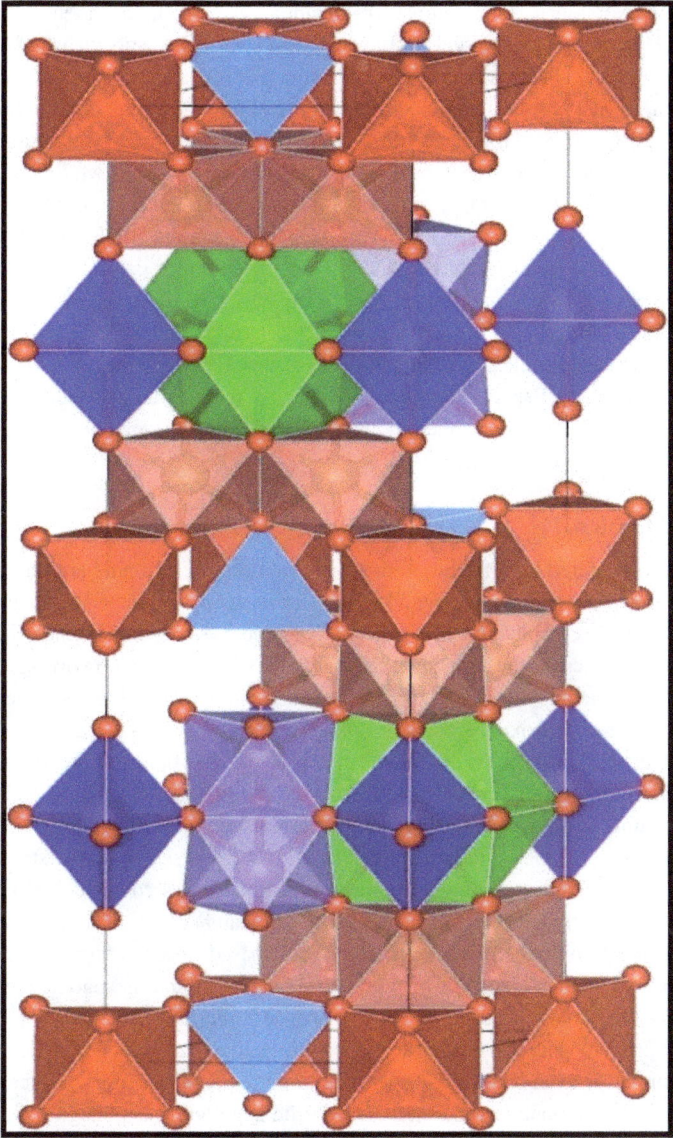

Figure 4. Presents the structure of hexagonal ferrites [8].

A range of techniques, for example co-precipitation [14], hydrothermal [15], sono-chemical [16], citrate [17], sol-gel [18], mechanical alloys [19], shockwaves [20], forced polyol hydrolysis [21] and even aqueous egg white [22], are available for nano-crystalline magnetic materials. Due to their specific qualities of chemical, mechanical and thermal stability, high coercivity, a high Curie anisotropic constant, moderate saturation, high electrical resistance and low eddy current loss, Nanosized $CoFe_2O_4$, $MnFe_2O_4$, $ZnFe_2O_4$, and $CuFe_2O_4$ NPs received impressionable attention during the past decade. The sol-gel and chemical co-precipitation procedures are outstanding approaches for the production of fine, homogeneous, nanostructured ferrites, among all methods examined. However, numerous innovative techniques make high-quality NPs cost-effective. While bulk ferrites remain a major category of magnet materials, the nanostructured ferrites examined are employed in a range of fields, including materials science, engineering, physics, chemistry, biology and medicine. Especially because of their low toxicity, their biocompatibility and their capacity to be easily controlled by magnetic field, these magetable ferrite NPs have shown excellent medical benefits.

4.   Hard ferrites

Ferrites are considered hard ferrites with strong coercivity. The main hard ferrites are the hexaferrites or the ferrite type M with hexagonal magnet plumbite or M structure, which are defined by the combination $MeFe_{12}O_{19}$ (Me=barium, strontium or lead). Their harsh magnet nature is because the hexagonal c axis is the favoured orientation owing to the huge magnetocrystalline analyzer. M-type ferrites are commonly utilised as permanent magnet materials, known as Ferroxdure, Koerox, Roboxid and Magnadure [23], among others. The expression "ferrite" must be regarded as ferrite of the type M unless expressly mentioned. The definition of hard ferrites as iron-based oxide compounds with substantial magnetic anisotropy implies that the majority of other hexages and cobalt ferrites must be included in the hard ferrites. Fe_3O_4-$CoFe_2O_4$ mixed crystals were based on the first commercial ceramic magnets, such as "Vectolite". The magnetic hardness comes out of an induced anisotropy, which progressively refrigerates at a temperature of 300 °C in a magnetic field, so the very anisotropic cobalt ions are ordered. Those based on cobalt ferrite were gradually discharged with the emergence of M-type ferrite magnets. Other hexagonal ferrites, such as $BaZn_2Fe_{12}O_{27}$ (ZnW), $BaZnFe_{14}O_{23}$ (ZnX) and $Ba_2ZnFe_{18}O_{30}$ (ZnU), also have uniaxial anisotropic property-like M-type ferrites. In principle, these compounds should thus create magnets similar or even better. However, due to more complex processing technologies, commercial development was missing. Hard ferrite is a good cheap magnet material that may be utilised up to 180 °C at temperatures. It has a wide range of applications due to its high magnetic and mechanical stability. For instance adhesive systems, speaker systems, electrical engines or sensor samples. It is necessary to push hard ferrite and sinter. The employment of grinding tools, diamond instruments, lasers and water jets makes a further machining possible. Hard ferrite magnets are non-toxic and do not contaminate the surroundings. Hard ferrite is extremely resistant to chemicals, like other ceramic materials. Concentrated inorganic acids are attacked primarily, based on temperature and length of contact.

Hard magnets are comprising of iron oxide and barium and strontium oxide. In order to produce the magnetic phase the raw materials are mixed together. The mixture is then crushed before sintering. In a magnetic field (an isotropic field) or without magnetic field (isotropic one), the resultant powder is pushed into one another (wet or dry), and finally sintered. Only through grinding are such procedures feasible. The cheapest form of magnet is accessible because of the low costs of raw material. Even in high external magnetic fields, hard ferrite magnets offer excellent electric insulation properties and are difficult to demagnetize.

4.1 Application of hard ferrites

4.1.1 Inductors

Ferrites are mainly utilised for inductive components in a wide range of electronic circuits, for example low noise amplifiers, filters, tension controlled oscillators, impedance matching networks. Among other trends, their current uses as inducers are subject to the overall trend of downsizing and integration as a ferrite multilayer for passive functioning electronic systems. The multi-layer technology is a vital technology in the mass manufacture of integrated devices, and multi-layer technology allows a high integration density. A few decades ago, multilayer condensers entered the market, and inducers began in the 80s. Furthermore, the ferrite film should be produced via a technique consistent with the integrated circuit manufacturing process to ensure high permeability at the working frequency. Sputtering gives a high degree of density for films, although it is sometimes difficult to precisely regulate the composition. Pulsed laser deposition leads to excellent quality films; yet it appears more easy to use and at a cheaper cost a technique involving ferrite film production by combining sol-gel and spin-coating [24].

4.1.2 Power

The power supply of a wide range of devices including computers, all sorts of peripherals, TV systems and videos and all forms of small and medium instruments dominate the power uses of ferrites. The major application is in systems called power supply for switching mode (SMPSs). The power signal of the mains is first corrected and then turned on at a high frequency to a ferrite transformer as regular pulses (usually rectangular), which is then ultimately revised to give the necessary power for the instrument. Increasing the operating frequency of the transformer might result in increased power supply and efficiency.

The recently adopted technique to improving the efficiency and resistivity of the ferrite cores is based on the reduction of eddy currents. In addition to the use of non-conducting additives which preferably lie on grain frontiers (and restrict the conductivity of the intergrain), MnZn and NiZn are coupled with $Mn_xNi_{0.5-x}Zn_{0.5}Fe_2O_4$ and produced using citrate precursor technique [25]. In the event of high-temperature power applications like some automobile power devices there is added challenge. The operating temperature for normal applications is increased from the customary 80–100 °C to 140 °C due to its near proximity to the motor.

4.1.3 EMI shielding

The substantial development of electronic equipment, including as high speed digital portfolios, digital cameras, scanners and more, in smaller spaces, has greatly increased the potential of electromagnetic interference upsetting each other (EMI). In particular, the rapid expansion of wireless communication has led to electrical and magnetic field interference. The performance deterioration of an electronic system induced by electromagnetic disturbance can be described as electromagnetic interference [26]. Noise generated on electric equipment is often higher than circuit signals at frequencies. The suppressors should be used in low-pass, circuits blocking signalling frequencies greater than the frequency-value, in order to avoid or at least minimise EMI. EMI repressors are built in a number of approaches: ferromagnetic metals [27], composites of ferromagnetic metal/hexaferrite [28], magnetic encapsulates [29], and carbon nanotube composites [30]. For decades, ferrite components were utilised for EMI suppressors. However, the downsizing trends and a growth in integration density have created specific requirements for these materials in recent years.

5.   Hard ferrites for memory devices

From a practical and basic point of view, ferrites are significant and intriguing materials. The nanosized $CoFe_2O_4$, $MnFe_2O_4$, $ZnFe_2O_4$, and $CuFe_2O_4$ were of particular interest because of their chemical and thermal stability and their unmistakable structural, optical, magnetic, electric and dielectric properties and the widespread technological potential of photoluminism, photocatalysis, wetness sensors, biosensors, catalysis, magnetic medication delivery, permaceticals, etc. among various ferrite products (hyperthermia). The standard inorganic-based flash memory instruments are manufactured on a limited number of rigid substrates utilising costly manufacturing techniques requiring high temperatures and vacuum conditions. However, there are growing demands for flexible storage components which may be produced on flexible and stretchy substrates utilising low-cost methods. Organic memory instruments that are cheap to manufacture, lightweight and supported by flexible plastic substrates received a lot of interest. A wide ranging effort was undertaken to enhance memory performance utilising various floating gate materials, such as polymer electres, organic semi-conductors and nanostructured materials, to implement flexibility-based flash memory devices. The conduct of a particular Gate–source biazard voltage depending upon the preceding operation is observed in floating memory gate systems with electrically biblical behaviours. The load carries produced by an external gate voltage at the half-conductor–dielectric interface can be picked up onto floating gates and released during erasure, leading to a threshold voltage change.

The transistor-type memory devices that employ nanostructured materials as loading sites for the trap are nanofloatting Gate (NFGM). Reciently, their outstanding performance, multi-leveled programming capabilities and platform compatibility for integrated circuits have garnered a significant amount of interest. In this case, unique NFGM equipment was manufactured by Jung et al., (2015) employing nanoparticles (NPs, $CoFe_2O_4$) as a load-trap site and pentacene as p-type semiconductor [31]. NP-based memory devices with $CoFe_2O_4$ show a large (Altogether 73.84 V) memory window, a high read-on/off ratio of ~2.98 x10^3 and good data retention. The extraordinary load-trap-/release capabilities of $CoFe_2O_4$ NPs around the oleate layer allow quick switching behaviours to be seen as an alternative tunnelling dielectric layer that facilitates the procedure of making the device. In addition, NFGM devices demonstrate great thermal stability and high mechanical and electrical stability in flexible storage instruments manufactured on plastic substrates.

Two order parameters contend with one form of order that suppresses the other in a hybrid supranational magnetic unit. New efforts to harness super-conductive and magnetic characteristics concurrently to produce new switching components with these two competing orders have been stimulated by recent interest in ultra-low power cryogenic memorys. Baek et al., (2014) presented a two-layer magnet spin valve incorporated in a Josephson junction that is programmable [32]. Their observations differentiate the suppression from the suppression of stray magnetic field in the supranational coupling owing to the magnetic layers exchange field which causes the supercurrent to be depairing. The removal of the superconducting order parameter from the exchange field is a controllable and switchable behaviour which can be measured also to nanoscale dimensions. These devices indicate that Josephson coupling is not volatile, independent in size, in magnitude and phase and can allow practical nanoscale superconducting storage systems.

Laser pulses are the shortest known stimulus for control of material magnetization and switching picoseconds magnetic devices to femtosecond time scales. Femtosecond laser-pulses have caused the rapidest changes in the magnetic state of matter, such that these pulses can result in magnetic data storage and memory technology with enhanced speed and energy efficiency. Materials for optical control of magnetism have been shown in the last decade and device ideas using such optomagnetic phenomena. In their study, Kimel et al., (2019) explored the ultrafast all-optical magnetic (AOS) switching as the most inefficient and quickest technique of magnetic writing [33]. Three major recording criteria are used in the physical processes responsible for AOS mechanisms to define materials appropriate for magnetic optic control, and to evaluate these mechanisms and materials: speed, concomitant dissipation and scalability. In particular, light-based switching magnetization outstrips other techniques with respect to the speed of the magnetic write-read recording event and the never previously low thermal charge (<6 Jcm^{-3}). Finally, integration of AOS into spintronic devices and a broad-ranging integration view towards random magnetic access storage and other low-energy memory applications.

The spin orbit torque (SOT) switching system is an energy-efficient way of controlling magnetization on magnetic memory instruments. The deterministic change in perpendicular memory bits, however, generally requires a further bias field in order to break lateral symmetry. Wang et al., (2018), employing the stress-mediated SOT switching technique, proposed a field-free deterministic perpendicular switching strategy [34]. The magneto-elastic stress-induced anisotrophy disrupts the lateral symmetry and controls the subsequent breakdown of the symmetry. The numerically simulating the stress-mediated SOT switching process is employed with a finite element model and a macro-spin model.

Magnetic nanostructures have broad use on a number of memory devices, with flux-closure state or with one domain state. However, it is seldom observed but still required that modulation of these variable states takes place inside one particular magnetic material. In this context, the modulation of $CoFe_2O_4$ building blocks assemblies was examined in these micromagnetic configurations in prototypic cobalt ferrite ($CoFe_2O_4$) nanostructures in various sizes and dimensions by Zhang et al., (2019) [35]. They found a multidominal structure of the $CoFe_2O_4$ nano-domain when the diameter is approximately 90 nm, where the Domain Walls (DW's) preferably are at the grain border and when the diameter is decreased, may be converted into a single domain state. Alternatively, when a $CoFe_2O_4$ nanostructure transitions from NW into the nanosheet (NS) the position of the DWs depends on the overall form of the NS is acquired by a flux closure domain status. Moreover, they establish that the magnet anisotropy and the magnetostatic energy are two major variables in crystallographic analysis and micromagnetic simulations impacting the micromagnet configuration in $CoFe_2O_4$ nanostructurings. Their test findings and simulations indicate that the modelling and dimensional modification of the magnetic nanostructures are effective to adjust the micromagnetic configuration.

In recent years, reserchers intensively studied the impact of interactions on magnetic characteristics of dense magnetic nanoparticle arrays. The magnetization and hysteresis and other magnetic measurements of zero fields with cooled fields (ZFC-FC) generally demonstrate that the presence of dipolar interactions in such systems increases energy barriers, and lead to greater blocking temperatures. The temperature blocking is the shift to the super-paramagnetic condition, in which the whole nanomagnetic particle magnet is subjected to the thermal fluctuation of a large paramagnetic spin. However, because of the random anisotropy barriers, spatial disturbance, and the presence of strong dipolar interactions, which could be either ferromagnetic or antiferromagnetic, magnetic nanoparticles with their inherent disorder and frustration have all the ingredients that are typical for atomic glass. The interacting magnetic moments are significantly longer, relaxation period is generally longer, and (dipolar) interactions are lengthy in nature. They are distinct from atomic spin glasses. Experimental evidence of slow dynamics and ageing, non-linear susceptibility divergence. The temperature dependent magnetization memory phenomenon, when the spin system reflects the FC magnetization curve thermal history throughout the heating cycle, everything seems to point out that the low temperature (super) spin-like glass-like stage exists in and refers there to concentrating nanoparticles. The source of memory effects, in particular, was discussed in the context of the involvement of polydispersion, dipolar interactions and RKKY interactions.

Self-assembled, high-density magnetic nanoparticles intriguing with prospective uses, such as nanotechnology, sensors and medicinal products, have garnered considerable research throughout the years. Controlled self-mounting of magnetic nanoparticles is a chemical synthesis technique since self-mounting does not match the lowest energy level in thermodynamics. The self-assembly of the nanostructure requires an input of external energy to control the self-assembly.In their study, Kumari et al., (2016) described a fresh technique by means of the supra-molecular lauric acid (anionic surfactant) as a structure that guides the hydro-thermic conditions of the manufacture of nanochristalline nickel ferrite of 5–9 nm particles [36]. A significant inter-particle interaction of nanoparticles as well as thorough ageing effect research clearly shows thermal, field, and temporal variations in magnetism. Interestingly, as the interparticle interaction of a nickel-ferrite nanoparticulary assembly the memory effect is found at low temperatures.

Since semi-permanent magnetization remained, memory cores were utilised for computer memory devices deployed aboard space shuttles as extremely dependable non-volatile memories.A unique thin film ferrite is used for a bubble memory. Digital information is saved on the basis of whether the cylindrical domain magnetization created from the film's surface up or down is up. Instead of a spinel-type ferrite (soft ferrite for transforming cores, etc) and a magneto plumbite-typical ferrite are utilised to produce a magnetic thin film for bubble memory (hard ferrite for ferrite magnets, etc.). Garnet-type ferrites are also used in circuit components such as circulators and isolators, which transmit a radio wave in one direction without backflow in communication equipment. Transformer core ferrites are utilised for spinel types, although different kinds of ferrites exist, including magnetoplumbites and garnets. They are employed in different fields where each one's benefit may be utilised.If intelligent species exist on a faraway planet in the cosmos, ferrites may have been produced as we did and amazing applications could have been created by them. Ferrites are electrical materials with endless potential, yet they are a substantial component of ordinary iron oxide.

The resistive performance switching features of $Pt/NiFe_{(2)}O_{(4)}/Pt$, for example low operating tension, high output of the device, long retention time (up to 10(5) s) and good endurance (up to 2.2 / 10(4) cycles), demonstrated the possibility of spinel ferrites in non-volatile memory device applications. The dominating mechanisms for conduction include Ohmic conduction in the low- and low-voltage regions of the high-resistance state and Schottky emissions in the high-voltage high-resistance area. Based on observations of temperature dependency in different resistance states for resistance and magnetic characteristics.

5.1 Multiple state memory devices

Electrical instruments which are based on the usage of magnetic oxides, and the characteristics of ferrites. Ferrite devices are classified into two groups, based on the magnetical softness or strength of the ferrite. Soft ferrites are mainly utilised in microwave devices, like transformers, inductors, and head recorders. Since Soft Ferrite usually shows 10^6–10^{11} times the electric resistivity of metals, ferrite components have significantly less losses in eddy current, and hence normally above 10 KHz at frequencies. Hard ferrite are employed as a storing media in magnetic devices and in permanent magnet motors, loudspeakers and holding devices.In research into multiferroic memories, discovery of a multi-state storage device based on LBMO is, of course, a milestone. However, a substantial memory impact on this system can only be seen at a very low temperature. Consequently, its application in the commercial memory device is limited.

Based on initial calculations, the $BaFe_{12}O_{19}$ M-Type shows frustrated anti-ferroelectricity with its bipyramidal Fe^{3+} trigonal sites. By introducing an external electric field to the antiferroelectric state, the ferroelectrical condition of $BaFe_{12}O_{19}$ may be rendered stable at room temperature by a suitable elemental replacement or strain engineering. M-type

Hexaferrite, as a novel multiferoic kind, offers the basis for the research of the frustrated antiferroelectricity phenomena and the development of multiple state memory systems.

Both ferroelectric and ferromagnetism are displayed by multiferroics. The unique phase combination of ferroelectricity and ferromagnetism allows for four physical polarisation states, two electric polarisation systems coupled with two magnetic polarisation states, and four state memory devices to be created. These memory devices can improve data storage capacity by a tenfold. Wu et al., (2012) synthesised $Sr_3Co_2Fe_{24}O_{41}$, a single phase with the effect of ferroelectricity at ambient temperature, ferro-magnetism and magneto-power copulation [37]. The data was written using electric and magnetic fields and read with a tiny magnetic field by magnetoelectric coefficient (αE).

The largest family of magnets presently utilised by bulk is made from permanent magnets based on hard hexaferritis. They produce a moderate induction of remanence but have important advantages in terms of accessibility, costs, corrosion resistance and the lack of the eddy current losses. Consequently, in chosen applications, ferrites are the most obvious option to substitute rare earths that do not require the greatest performance magnets. The door to a bigger substitution might be opened if the remanence of a ferrite-based magnet could be somewhat enhanced, as shown in Fig. 5.

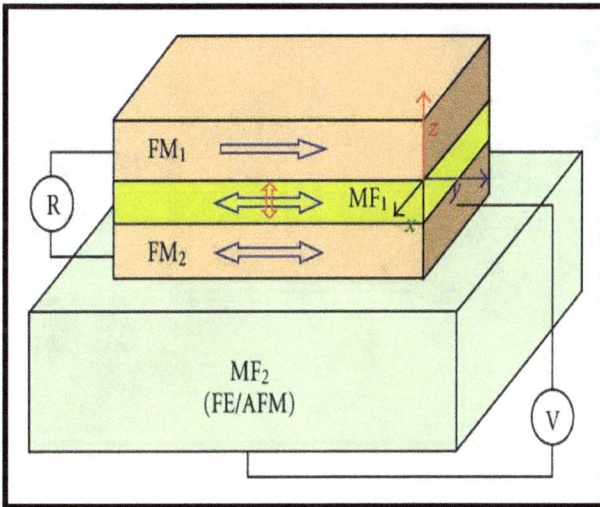

Figure 5. Presents the image of multiple state memory device [37].

While traditional (charge-based) memories in diverse forms fit their functions in computers and other electronic devices, defects in their characteristics lead to active research into alternative and emerging memories. In particular, it has been hard to achieve at the same time conflicting non-volatility criteria and rapid low-voltage (low-energy) switching. Here we have a floating memory cell with oxide free based on III-V heterostructure semiconductor and a juncture-free channel, which reads stored data without destruction [39], as shown in Fig. 6. The exceptional 2.1 eV conduction band offset of InAs/AlSb and a three barrier Resonant tunnel construction provide the non-volatile data retention of at least 104 s in conjunction with a switching of 0.45 V. The combination of low voltage and small capacity involves 100 times and 1000 times less switching electricity per unit area than the dynamic random access memory and flash drives. The gadget can therefore be regarded a novel and prospective emerging memory.

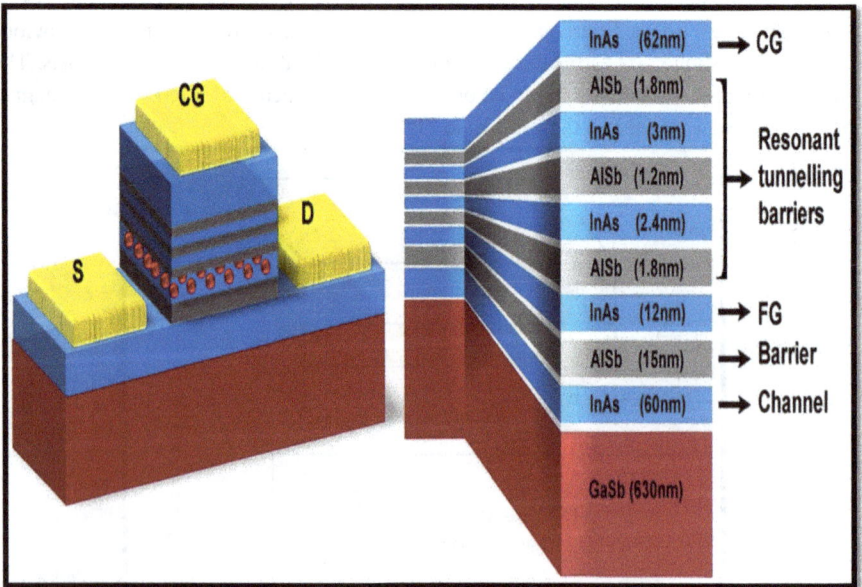

Figure 6. Presents the image of multiple state memory device [39].

5.2   Magnetic core memory

Even as the computer era was dawning, magnetic core memory contributed to facilitate data storage and demonstrated surprising staying power in an area that continuously replaces new and better components. Critical components often travel from cutting-edge to outdated in a short period in the fast developing computer world. However, there are situations when a critical mechanism shows extended power. Magnetic core memory is one example of this. It contains a small round ceramic (called cores), loaded with data, and is connected by a web of wires as one of the initial phases of random access computer memory. This storage system is usually called "core" as shown in Fig. 7.

Figure 7. Magnetic Core Ferrite Memory (1940) [40].

The mechanism of operating magnetic memory is based on a feature that affects all ferromagnetic components. These may be magnetised in two permanent states. The two magnetic states around the circle are identified for the ferrite ring, both counterclock and clockwise. The two conducting cables must pass through the magnetization core state.

A conducting wiring creates a magnetic field and it is possible to induce the magnetise state of the kernel to change the strength and direction of the current passing through it, generating what is described as the hysteresis cycle.

Two US scientists who immigrated from China credit its existence to the core memory. The development of the pulse transfer control device was accredited in 1949 by Way-Dong Woo and An Wang. The name refers to how the electromagnetic effect of cores governs electricity transmission into electromechanical structures. Two major improvements expedited the route to the modern computer within two years following Wang and Woo's work. First, following a reading round, Wang designed the writing to avoid deleting the words as they were being read. Next came the development of the coincidence-current system, which permitted the operation of huge groups of cores by far less wires than had previously been feasible. The method required a 45-grade corner wire to be placed in the buildings, which could simply not be done with a machine. Workers with exceptional hand-eye coordination have, thus, done the task with microscopes.

Over a decade, the manufacturing core of industrial facilities was in Asia. The price of the core dropped to a level that made it the universal option of memory, displacing other more expensive systems with low salaries and cheap overheads. As the core memory developed, its costs continued to decline from roughly one dollar a bit to about $0.01 a bit. Even if the core was already in operation for some years, Wang did not patented it until the mid-1950s. Following a flourishing litigation, IBM settled numerous millions of dollars with Wang and took up the patent in its entirety. At its heyday, the core of memory was one of the most respected segments of a set of comparable technologies that employed different computer-related operations using magnetic trends from different resources. At the same time, the electronics from vacuum tubes were progressed, but didn't survive long enough. Magnetic devices, on the other hand, featured many of the greatest properties of components from solid state that would later be replaced.

Because power was not required for its content to be kept, core memory was less costly and easier to store. In addition, because to its usage in military equipment and in vehicles it is unavoidable to the effects of radiation. The initial generation of computers on the Space Shuttle employed core memory, which proved its resilience even after the Challenger disaster in 1986. However, the core memory has been steadily supplanted by semiconductors and has continued to disappear from the high-tech industry.

Conclusion

For memory devices, ferrites are the key material. In this article, we examine in depth ferrites including their categorization and preparation technique. We also provide a short introduction to hard ferrites and their use in memory devices. Ferrites are materials with diverse electrical and magnetic characteristics that are useful for different purposes. It should be noted that spin transmission torques can alter the magnetization by using a Magnetic Tunnel Interface, followed by a random access memory of the spin transfer torque magnet as a switching mechanism. Also, it can be noticed that nanostructured materials are nanofloatting gate using the transistor memory devices that load the trap (NFGM). This chapter therefore offers the memory devices a path forward.

The main kind of electronic MEMORY utilised before semiconductor memory chips are invoked on the computer. It included hundreds of small rings called core, each of which was manufactured from FERRITE magnet material and stitched on three fine copper wires. The entire was like a knitted textile, created a two-dimensional mesh. When the two proper vertical and horizontal wires were sent, a particular core could be addressed and either read or changed its direction of magnetization using a third wire, thereby acting as a memory RANDOM ACCESS. The size of its components reduced thousands of bytes instead of millions of the amount of core memory that could fit into a computer.

Acknowledgement

Authors would like to acknowledge the chancellor of Shoolini University for his continuous motivation.

References

[1] W. H. Bragg, The structure of the spinel group of crystals, Phil. Magz. 30 (1915) 305-315. https://doi.org/10.1080/14786440808635400

[2] R. Valenzuela, Novel Applications of ferrites, Phys. Res. Int. (2012) 1-9. https://doi.org/10.1155/2012/591839

[3] F. Bertaut, F. Forrat, Structure des ferrites ferrimagnetiques des terres rares, Compt. Ren. de l'Acad. des Sci. 242 (1956) 382-384.

[4] S. Geller M. A. Gilleo, The crystal strutucre and ferromagnetism of Yttrium-iron garnet Y3Fe2(FeO4)5, J. Phys. Chem. Solids. 3 (1957) 30-36. https://doi.org/10.1016/0022-3697(57)90044-6

[5] A.K.Singh,M.G.H.Zaidi,P. L. Sah, R. Saxena, A Review on Classification of Various Ferrite Particles on the Basis of Crystal Structure, Eur. J, Adv. Eng. Tech.5 (2018)350-354.

[6] R. Verma, A. Chauhan, K. M. Batoo, R. Jasrotia, A. Sharma, R. Kumar, M. Hadi, E. H. Raslan, J. P. Labis, A. Imran, Review-Modulation of Dielectric, Ferroelectric and Piezoelectric properties of Lead-free BCZT ceramics by doping, ECS J. Solid State Sci. Tech. 10 (2021) 073004. https://doi.org/10.1149/2162-8777/ac0e0d

[7] G. Albanese, M. Carbucicchio, A. Deriu. Substitution of Fe3+ by Al3+ in the trigonal sies of M-type hexagonal ferrites, Il Nuovo Cimen. B. 15 (1975) 147-158. https://doi.org/10.1007/BF02894778

[8] C. De J. Fernandez, C. Sangregorio, J. De la Figuera, B. Belec, D. Makovec, A. Quesada. Progress and prospects of hard hexaferrites for permanent magnet, J. Phys. D: Appl. Phys. 54 (2021) 153001. https://doi.org/10.1088/1361-6463/abd272

[9] M. Abe, Y. Kitanoto, K. Matsumoto, M. Zhang, P. Li. Ultrasound enhanced ferrite plating bringing breakthrough in ferrite coating synthesised from aqueous solution, IEEE Trans. Magn. 3 (1997) 3649-3651. https://doi.org/10.1109/20.619526

[10] J. J. Cuomo, R. J. Gambino, J. M. E. Harper, J. D. Kuptsis, J. C. Webber, Significance of negative ion formation in sputtering and sims analysis. J Vac. Sci. Tech. 15 (1978) 281-287. https://doi.org/10.1116/1.569571

[11] D. Dijkkamp, T. Venkatesan, X. D. Wu, S. A. Shaheen, N. Jisrawi, Y. H. Min-Lee, W. L. Melean, M. Croft, Preparation of Y-Ba-Cu oxide superconductor thin films using pulsed laser evaporation from high Tc bulk material, Appl. Phys. Lett. 51 (1987) 619. https://doi.org/10.1063/1.98366

[12] D. M. Lind, S. D. Berry, G. Chern, H. Mathias, L. R. Testardi, Growth and structural characterization of Fe3O4 and NiO thin films and superlattices grown by oxygen plasma as sintered molecular beam epitaxy. Phys. Rev. B. 45 (1992) 1838-1850. https://doi.org/10.1103/PhysRevB.45.1838

[13] Y. Suzuki, R. B. Van Dover, E. M. Gyorgy, J. M. Phillips, R. J. Felder, Exchange coupling in single-crystalline spinel structure (Mn,Zn)Fe2O4/CoFe2O4 bilayers. Phys. Rev. B.53 (1996) 14016-14019. https://doi.org/10.1103/PhysRevB.53.14016

[14] J. M. Yang, W. J. Tsuo, F. S. Yen, Preparation of ultrafine nickel ferrite powders using mixed Ni and Fe tartrates, J. Solid State Chem. 145 (1999) 50-57. https://doi.org/10.1006/jssc.1999.8215

[15] J. Zhou, J. Ma, C. Sun. Low-temperature synthesis of NiFe2O4 by a hydrothermal method. J. Am. Ceram. Soc. 88 (2005) 3535-3537. https://doi.org/10.1111/j.1551-2916.2005.00629.x

[16] K. V. P. M. Shafi, Y. Koltypin, A. Gedanken, Sonochemical preparation of nanosized amorphous NiFe2O4 particles, J. Phys. Chem. B. 101 (1997) 6409- 6414. https://doi.org/10.1021/jp970893q

[17] S. Prasad, N. S. Gajbhiye, Magnetic studies of nanosized nickel ferrite particles synthesized by the citrate precursor technique, J. Alloy. Compd. 265 (1998) 87-92. https://doi.org/10.1016/S0925-8388(97)00431-3

[18] D. -H. Chen, X. -R. He, Synthesis of nickel ferrite nanoparticles by sol-gel method, Mater. Res. Bull. 36 (2001) 1369-1377. https://doi.org/10.1016/S0025-5408(01)00620-1

[19] Y. Shi, J. Ding, X. Liu, J. Wang, NiFe2O4 ultrafine particles prepared by co-precipitation/mechanical alloying, J. Mag. Mag. Matr. 205 (1999) 249- 254. https://doi.org/10.1016/S0304-8853(99)00504-1

[20] A. Kale, S. Gubbala, R. D. K. Misra, Magnetic behavior of nanocrystalline nickel ferrite synthesized by the reverse micelle technique, J. Mag. Mag. Matr. 277 (2004) 350-358. https://doi.org/10.1016/j.jmmm.2003.11.015

[21] Z. Beji, T. Ben Chaabane, L. S. Smiri, Synthesis of nickelzinc ferrite nanoparticles in polyol: morphological, structural and magnetic studies, Phys. Stat. Solidi A. 203 (2006) 504-512. https://doi.org/10.1002/pssa.200521454

[22] S. Maensiri, C. Masingboon, B. Boonchom, S. Seraphin, A simple route to synthesize nickel ferrite (NiFe2O4) nanoparticles using egg white, Scripta Matr.56 (2007) 797-800. https://doi.org/10.1016/j.scriptamat.2006.09.033

[23] H. Stäblein, Hard ferrites and plastoferrites, In: Wohlfarth.7 (1982) 441-602. https://doi.org/10.1016/S1574-9304(05)80093-8

[24] C. Yang, F. Liu, T. Ren, Fully integrated ferrite-based inductors for RF Ics. Sens. Acrt. A. 130-131 (2006) 365-370. https://doi.org/10.1016/j.sna.2005.10.024

[25] A. Verma, M. I. Alam, R. Chatterjee, T. C. Goel, R. G. Mendiratta, Development of a new soft ferrite core for power applications, J. Mag. Mag. Matr.300 (2006) 500-505. https://doi.org/10.1016/j.jmmm.2005.05.040

[26] G. Stojanovic, M. Damnjanovic, V. Desnica, High-performance zig-zag and meander inductors embedded in ferrite material, J. Mag. Mag. Matr.297 (2006) 76-83. https://doi.org/10.1016/j.jmmm.2005.02.058

[27] Y. B. Feng, T. Qiu, C. Y. Shen, X. -Y. Li, Electromagnetic and absorption properties of carbonyl iron/rubber radar absorbing materials, IEEE Trans. Magn. 42 (2006) 363-368. https://doi.org/10.1109/TMAG.2005.862763

[28] B. W. Li, Y. Shen, Z.-X. Yue, C.-W. Nan, Enhanced microwave absorption in nickel/hexagonal-ferrite/polymer composites, Appl. Phys. Lett. 89 (2006) 132504. https://doi.org/10.1063/1.2357565

[29]R. C. Che, C. Y. Zhi, C. Y. Liang, X. G. Zhou, Fabrication and microwave absorption of carbon nanotubes CoFe2O4 spinel nanocomposite, Appl. Phys. Lett.88 (2006) 1-3. https://doi.org/10.1063/1.2165276

[30]C. Xiang, Y. Pan, X. Liu, X. Sun, X. Shi, J. Guo, Microwave attenuation of multiwalled carbon nanotube-fused silica composites, Appl. Phys. Lett.87 (2005) 1-3. https://doi.org/10.1063/1.2051806

[31]. J. H. Jhung, S. Kim, H. Kim, J. Park, J. H. Oh, High performance flexible organic nano-floating gate memory devices functionalised with cobalt ferrite nanoparticles, Small 11 (2015) 4976-4984. https://doi.org/10.1002/smll.201501382

[32] B. Baek, W. H. Rippard, S. P. Benz, S. E. Russek, P. D. Dresselhaus, Hybrid superconducting magnetic memory device using competing order parameters, Nat. Comm. 5 (2014) 3888. https://doi.org/10.1038/ncomms4888

[33] A. V. Limel, M. Li, Writing magnetic memory with ultrashort light pulses, Nat. Rev. Matr. 4 (2019) 189-200. https://doi.org/10.1038/s41578-019-0086-3

[34] Q. Wang, J. Domann, G. Yu, A. Barra, L. Wang G. P. Caman, Strain-Mediated Spin-Orbit-Torque Switching for Magnetic Memory, Phys.Rev. Appl. 10 (2018) 034052. https://doi.org/10.1103/PhysRevApplied.10.034052

[35] J. Zhang, S. Zhu, W. Xia, J. Ming, F. Li, J. Fu, Micromagnetic Configuration of Variable Nanostructured Cobalt Ferrite: Modulating and Simulations toward Memory Devices, ACS Appl. Mater. Interfaces 11 (2019) 28442-28448. https://doi.org/10.1021/acsami.9b07502

[36] V. Kumari, K. Dey, S. Giri, A. Bhaumik, Magnetic memory effect in self-assembly nickel ferrite nanoparticles having mesoscopic void spaces, RSC Adv. 6 (2016) 45701-45707. https://doi.org/10.1039/C6RA05483H

[37] A. Roy, R. Gupta, A. Garg, Multiferric magnetoelectric composites and their applications, Adv. Cond. Mater. Phys. (2012) 926290.

[38] J. Wu, Z. Shi, J. Xu, N. Li, Z. Zhang, H. Heng, Z. Xie, L. Zheng, Synthesis and room temperature four state memory prototype of Sr3Co2Fe24O41 multiferroics, Appl. Phys. Lett. 101 (2012) 122903. https://doi.org/10.1063/1.4753973

[39] O. Tizno, A. R. J. Mashall, N. F. Delgado, M. Herrera, S. I. Moling, M. Hayne, Room temperature operation of low voltage, non-volatile, compound semiconductor memory cells, Sci. Report 9 (2019) 8950. https://doi.org/10.1038/s41598-019-45370-1

[40] B. North, O. Nash, Magnetic core memory reborn, (2011).

Keyword Index

About the Editors

Dr. Gagan Kumar Bhargava obtained his Ph.D. degree in Physics discipline from Himachal Pradesh University, Shimla, India and is working as Professor at the Department of Physics, Chandigarh University, Gharuan, Mohali, Punjab, India. His current research interests are synthesis of nanomaterials for antenna applications, EMI shielding applications, electromagnet applications and cancer treatment. He has more than 15 years of teaching and research experience. His current research interests are in the field of nanoferrites and multiferroic materials. He has published more than 50 research papers in different peer reviewed international journals including 6 book chapters. He has guided 3 Ph.D's, 1 M. Phil and 12 M. Sc. dissertations.

Dr. Pankaj Sharma received his Ph.D. degree in Physics from JUIT, Waknaghat, India and is serving as Professor in the Applied Science Department, NITTTR Chandigarh. Prior to joining NITTTR in January 2020. He has worked for more than 12 years in the Department of Physics & Materials Science, Jaypee University of Information Technology, Waknaghat under various positions. He is an experimentalist by inclination and his research interests include the development of materials in 0-D, 1-D, and 2-D for various optoelectronic, thermoelectric and biomedical applications. Few materials of recent interest include ferrites, chalcogenides, polymers, and dilute magnetic semiconductors. He has published more than 125 research articles in SCI/Scopus index journals, among these 100 research papers in SCI indexed journals. He has published 5 book chapters and 4 review papers. He has contributed more than 50 articles to various conferences. Dr Sharma has delivered more than 50 expert lectures in conferences/workshops/faculty development programmes etc. He has executed 02 government sponsored projects. He has successfully supervised 5 research scholars to earn their PhD degree. He is senior member of IEEE, founder member & joint secretary of Materials Research Society of India- Himachal Pradesh Chapter. He is also serving as member of editorial board of two international journals.

Dr. Sumit Bhardwaj is currently working as an Assistant Professor at the Department of Physics, Chandigarh University, Gharuan, Mohali, Punjab, India. He has more than 11 years of research and teaching experience. His current research interests are in the field of multiferroic materials and Polymer nanocomposites. He has published more than 16research papers in different peer reviewed international journals and 3 book chapters. He has guided 2 M.Tech. students and 22 under graduate students for their dissertations and projects.

Dr. Indu Sharma is currently working as an Associate Professor at the Department of Physics, Career Point University, Himachal Pradesh, India. She has obtained her Ph.D in

Physics from National Institute of Technology, Hamirpur, India and has more than 15 years of research experience. She has published more than 20 research papers in different peer reviewed international journals and 5 book chapters. She has guided 1 Ph.D, 3 M. Phil and 30 M. Sc. dissertations.

www.ingramcontent.com/pod-product-compliance
Lightning Source LLC
Chambersburg PA
CBHW071208210326
41597CB00016B/1726